Empty Brain
Happy Brain

Niels Birbaumer is a psychologist and neurobiologist. He is a leading figure in the development of brain–computer interfaces, a field he has researched for 40 years, with a focus on treating brain disturbances. He has been awarded numerous international honours and prizes, including the Gottfried Wilhelm Leibniz Prize and the Albert Einstein World Award of Science. Professor Birbaumer is co-director of the Institute of Behavioural Neurobiology at the University of Tübingen in Germany, and senior researcher at the Wyss Centre for Bio- and Neuro-engineering in Switzerland.

Jörg Zittlau is a freelance journalist, and writes about science, psychology, and philosophy, among other topics. He is also the author of several bestsellers.

Empty Brain
Happy Brain

how thinking is overrated

Niels Birbaumer and Jörg Zittlau

Translated by David Shaw

SCRIBE

Melbourne • London

Scribe Publications
2 John St, Clerkenwell, London, WC1N 2ES, United Kingdom
3754 Pleasant Ave, Suite 100, Minneapolis, Minnesota 55409, USA

Originally published as *Denken wird überschätzt* in German
by Ullstein in 2016
First published in English as *Thinking is Overrated* by Scribe in 2018
This edition published 2019

Typeset in Adobe Caslon Pro 12/16 pt by the publishers

Printed and bound in the UK by CPI Group (UK) Ltd, Croydon CR0 4YY

Scribe Publications is committed to the sustainable use of natural resources
and the use of paper products made responsibly from those resources.

9781911344582 (UK edition)
9781947534322 (US edition)
9781925693454 (e-book)

A CiP data record for this title is available from the British Library.

scribepublications.co.uk
scribepublications.com

Contents

Introduction

A Parachute Jump into Emptiness

I was green around the gills. Only minutes earlier, I had been chattering away happily with a more or less healthy hue to my complexion as I boarded the plane — with a plan to demonstrate how our wireless technology for measuring heart and sweat-gland activity works under unusual conditions. But now, here I was, just about to jump out of the plane, with only a parachute to save me, and I had taken on the colour of a vampire who's been snacking on the wrong blood group. Later, this will even be recognisable on photos of the event.

My mouth was so dry that my tongue stuck to the roof of my mouth; my knees were so weak, they were trembling as I staggered towards the hatch. There was not a word to be heard from me now, not a peep. I would never have managed to put together a sensible sentence, anyway,

1

as my mind was racing, without contributing anything constructive to the situation.

My friend, the brain scientist and musician Valentino Braitenberg, described the brain as a 'thought pump', continually drawing things up from the deep. Right now, my 'pump' was just about to go into hyperactive collapse, unable to draw anything but snatches of thoughts up from the depths, like a shipwrecked sailor trying desperately to bail out her leaking lifeboat with an empty yoghurt pot.

Then, finally, I jumped. I suspect someone must have pushed me, but I have no recollection of it now. Just as I have generally very little memory of anything from the moment I jumped to the moment I landed. Suddenly, the panic within me disappeared. The carrousel of thoughts in my head stopped spinning and I was simply falling, with the sky above and the slowly approaching forests below me.

It was a moment of rapture — my 'self' no longer seemed to exist. The fear I had felt before jumping was gone, and it was not replaced by a new fear, because there was nothing I could do anyway. Our wireless-technology project was certainly no longer of any concern to me, and all my other day-to-day worries were swept up into the sky by the wind that was thundering in my ears. I've heard of mountaineers seeing their whole life flash before them as they plummet from the heights. But for me there was: nothing. Just emptiness.

The world was still there, but the borders between it and me became blurred. The others who jumped with me later told me I let out a yell for several seconds as I was falling, the like of which they had never heard emanating from me. I can't remember it. I don't even remember my parachute opening. All I can remember is landing, which in my case involved the branches of a tree and a few light injuries because I forgot to steer. And I remember my deep disappointment at the fact that it was over. I felt as if I had awoken from a wonderful dream but could not remember what made it so beautiful.

I have not done another parachute jump since. Not out of fear of the fall itself, which was appeased by that first jump. My fear is a different one. It is namely the fear that plummeting into the depths will never be as wonderful as it was the first time: so wonderfully *empty*.

What remains when we no longer think or feel?

Brain scientists don't usually have much truck with emptiness. Their work revolves around behaviours, thoughts, and emotions — their inadequacies, and also their potential.

We now know that our brain is an organ of enormous plasticity. It is always able to keep learning and adapting from our early youth to our old age. Infants grow up speaking two languages with no problem at all, old people can learn to juggle or play a musical instrument even in extreme old age, criminals can become useful members of society, and, contrariwise, successful business executives can become desperate criminals. The possibilities are many — both desirable and undesirable — and include the ability to cope with crisis situations. It is a constant source of amazement the way traumatised children, maltreated concentration-camp survivors, and victims of war somehow manage to lead fulfilled lives again. Other people, by contrast, fall into despair at nothing more tragic than a lost football match.

In all these cases, problem-solving thinking is what is required — and our thought pump begins working at full power. This not only brings us the realisation that the world exists, but also makes us realise that *we* exist in that world. René Descartes summed this up in his famous phrase: *cogito ergo sum* — *I think, therefore I am*. Everything may be uncertain and in doubt, but the fact remains that it is *I* who am thinking those doubts; and in the first instance, that sounds comforting.

In another way, however, it also sounds worrying, since it raises the question: what remains of us when we no longer think or feel anything? Are we then — nothing? Must we fear sinking into a sea of emptiness and eventually dissolving away?

In our daily lives at least, that fear does not appear to play an important role. We find it almost unbearable when the television breaks down or the internet is cut off, or when we have nothing to do or no one to be with. In a survey of young men and women, a third of the respondents said they would rather go without sex than their smartphone if they were marooned on a desert island. Other surveys have shown that people's fear of boredom is similar to their fear of cancer. Almost as if to say: better to be fatally ill than empty. Yet another study found that healthy volunteers with no masochistic tendencies would rather give themselves harmless but unpleasant electric shocks than sit and wait for 15 minutes (see Chapter 1).

The fear of emptiness also plays a major role in many medical conditions (see Chapter 10). For example: dementia, which eventually leads to complete apathy. Or borderline personality disorder and depression, which lead patients repeatedly to express the lack of meaning and the pointlessness of their existence. Psychopaths and adults with attention-deficit hyperactivity disorder (ADHD) are driven to their abnormal behaviours by their fear of emptiness. They need powerful stimuli to escape it, which is why they torment animals and people, risk huge sums on the stock markets, or speed down the motorway at 200 kilometres per hour.

A study carried out at the University of Innsbruck in Austria showed that people with aggressive, sadistic, or psychopathic behavioural traits have a great preference for bitter-tasting foodstuffs.[1] The reason for this is that bitterness is one of the extreme, even potentially life-threatening stimuli that psychopaths need. Many poisonous substances taste bitter, and that is why stimulating the bitterness receptors on the tongue sends the brain into alarm mode. Thus, black coffee and gin and tonic are among the kicks that psychopaths need in their lives. It's no accident that James Bond drinks extremely dry vodka martinis.

In the experience-driven society we live in, the extent of our fear of emptiness can be seen in the fact that almost 30 per cent of people in Germany have signed a living will. Such a document determines that life-prolonging measures should be terminated if the patient is left bedridden and completely paralysed, with no hope of recovery. People's fear of this state of absolute inactivity is so great that they would rather be dead. However, very few people know what life might be like for them when they have lost the ability to do anything.

At the University of Tübingen's Institute of Behavioural Neurobiology, we have spent many years working on establishing contact with completely paralysed and locked-in patients (see Chapter 11). We have not only achieved various degrees of success in this, but have also been able to ascertain that these people appear to enjoy a high quality of life. For some, even higher than that of healthy people! This despite the fact that they were no longer able to move a single muscle, and their brains showed mainly low-frequency activity, which could be described as typical of 'running on empty'.

Or is the very reason for their happiness *because* their lives are 'filled' with emptiness?

Emptiness provides an unfettered view of the world
Some philosophers even postulate that emptiness is the source of a special kind of existential happiness (see Chapter 2). For example: Gautama Buddha, or Arthur Schopenhauer, who saw the *will* as the source of all suffering because it always makes us desire and do things without ever leading to final satisfaction. Better to find a way of extinguishing it. This may be through compassion, because it detracts our attention away from our own will, or through meditation, because it helps to offer a view which is free of desire.

Or — according to Schopenhauer — through music, because it is a direct and immediate copy of the will, allowing our individual wills to merge with it and find peace. Brain research has, in fact, found scientific evidence for this theory (see Chapter 9). In Tübingen, we were able to show that music with a strong rhythmical beat in particular produces simple, i.e. mathematically predictable and therefore calculable neuro-electrical oscillation patterns in the brain, with only slight irregularities. Blues and techno music thus offer a better way to emptiness than classical music or free jazz.

Researchers have also discovered that our brains work in various different emptiness states, and prefer the so-called 'twilight state', in which neurons fire off in the low-frequency waveband and the thalamus closes its gates to limit the stimuli that can reach the upper levels of the brain (see Chapter 3). Thus, the brain has been proven to have an emptiness mechanism. The most fascinating aspect of this is that the brain very much likes to switch it on, as evidenced by the fact that such states recur repeatedly throughout the day, and especially at night as we sleep. We are 'pro-emptiness'. As much as emptiness sometimes instils fear in us, it also attracts us. And this is astonishing, since it offers nothing — no concrete reward — that might cause such a preference in the brain. As a consequence, we must ask: what is it that we get from emptiness that makes us seek it out?

Closer inspection reveals an astonishing number of answers to this question. For example, emptiness allows our defence systems to take a rest (see Chapter

4). These are located mainly in the deeper regions of the brain, and their job is to identify sources of danger as early as possible, which is why the human species would undoubtedly never have survived without them. On the other hand, they also give our brains a natural 'catastrophic bent', as the psychologist Martin Seligman has so aptly put it: we tend to see danger all around us. And in a world such as ours, with all its complexities and the many potential dangers they occasion, this means our thought pumps are constantly concerned with averting or avoiding danger. Our defence systems are basically on permanent high alert. This is energy-sapping and — as psychosomatics frequently stress — opens the way for many illnesses. Emptiness can offer respite and relief from this. It helps put things into perspective, making them seem less problematic.

But that's not all. Emptiness can also create new stimuli. This might sound absurd at first: surely nothingness cannot create anything? But when our brain activity forms a gently lapping ocean of low-frequency waves, high-frequency attention waves stand out more easily. If we place people in a floatation tank, where not only their senses of hearing, sight, touch, and taste are shut down, but, most importantly, also their proprioception — that is, their spatial sense of their own body — they feel blissfully happy and profoundly relaxed. Some even report having new, creative ideas in this state of 'sense-lessness' (see Chapter 6).

We see similar phenomena in connection with meditation: a brainwave-sea of emptiness, from which occasional rocks of absolute but disinterested attention stand out. It should be noted, however, that the forms of a meditation practitioner's brainwaves depend heavily on how far he or she is able to descend into a truly meditative state. Among followers of the Indian guru, and founder of the Transcendental Meditation technique, Maharishi, we found a relatively large number of meditators who simply fell into a state of sitting slumber. We then went on to examine followers of Zen meditation. One of its main proponents in the USA did at least remain awake, but his brainwaves also showed nothing that is not normally found in everyday life.

It was only the 'original' Zen practitioners, from Asia, whose brain activity showed that they were neither asleep nor awake in the everyday sense. These meditation experts detached the front part of their brains from the rear, thus also severing the link between their sensory perceptions and the meaning of those perceptions. In other words, they were able to render the world empty of meaning and observe it as it really is, in a dispassionate, functionless, and objective way (see Chapter 7).

There is no alternative: emptiness requires trust

There are many ways to achieve emptiness. Apart from meditation, floatation tanks, music, and dance, these ways also include sex, religion, and epilepsy — three things with quite a bit in common (see Chapter 8). And there are probably many more.

While writing this book, I was once again aided by the philosopher, science journalist, and — particularly helpfully this time — experienced musician Jörg Zittlau, and, during the process, new potential techniques to achieve emptiness kept occurring to us.

One such example is art, to which Schopenhauer assigned a certain potential for release from will. Others include things such as cheering in the crowd at sporting events or marching in step, which might not be quite so culturally highbrow, but which have just as much of an 'emptying' effect on some people. Some sports enthusiasts enter a kind of 'emptiness zone' as they rock climb, row, or run a marathon; for other people, doing the ironing is enough to reach this state.

Some types of drugs also bring about such emptiness, but many have rather hefty side effects. I had a very intense experience of emptiness with curare (see Chapter 6), but this arrowhead poison used by the indigenous people of South America is famous for causing complete paralysis, and so cannot be used except in the presence of an experienced anaesthetist to ensure continued breathing. Which brings us to a pivotal point to be considered on the way towards emptiness.

I would never conduct an experiment that involved paralysing the respiratory system unless there were an anaesthetist present whom I trust implicitly. If that trust is not there, what remains is caution and fear — and those are barriers to achieving emptiness. This is not only true for those experimenting with curare. Anyone who makes only a half-hearted attempt to meditate or keeps one eye on the exit during a floatation-tank session will fail to achieve a state of emptiness. Mediocre musicians will be less able to lose themselves in the music than practised professionals who do not need to concentrate so hard on mastering their instrument. Completely locked-in patients achieve a higher level of satisfaction with life than many paraplegics, presumably because they have come to terms with their fate and their loss.

During my parachute jump, I only experienced a state of emptiness because there was nothing I could do about the situation I was in once I had jumped. Positive emptiness can only occur when we allow ourselves to surrender to a given situation completely, with trust, and without compromise or a feeling of regret for what we lose when we gain emptiness. We cannot have an alternative to emptiness in mind, or feel fearful of it, or hope to gain something from it. Otherwise, it will not work.

Some readers will now be asking themselves what on Earth we are talking about. What is this emptiness, which can only occur when fear, mistrust, regret, and expectation are banished? Jörg Zittlau and I have spent much time debating the definition of emptiness. We discovered many new aspects of emptiness — but no definition of it. Slower rhythms occur in the brain, the defence and stress systems in the brain are inhibited, a strange kind of openness occurs in the senses, thinking in words and sentences is rolled back, and any feeling of 'drive' ebbs away. The difficulty in defining this emptiness comes from the fact that it is a description of something that isn't there, and which lacks structure, form, content, meaning, and anything else that we use as aids to thinking. How can we define such a thing? Or is that perhaps the definition itself? We do not know. But anyone who sets out to write a book about emptiness must accept this lack of a definition. And I suppose the same is true of anyone who reads such a book.

1
There's Always Something

*how we have banished emptiness
from our lives*

'Herman, what are you doing in there?'
  'Nothing.'
  'Nothing? What do you mean, nothing?'
  'I'm not doing … anything.'
  'Nothing at all?'
  'No.'
  'Absolutely nothing?'
  'No. I'm just sitting here.'
  'You're just sitting there?'
  'Yes.'
  'But you must be doing something?'
  'No.'
  'Are you thinking about something?'
  'Nothing in particular.'

This scene with a married couple — she is busy in the kitchen and is annoyed that he is just sitting in his armchair in the living room — is one of the classic sketches by the German master of observational comedy, Loriot. Of course, it can be seen as an exaggerated depiction of the communication problems that occur between the sexes. But this dialogue also caricatures another issue: our inability to endure nothingness.

Herman's wife is clearly distressed. Not because she herself is yet again stuck working in the kitchen, but because her husband is just sitting there. Doing nothing. And as if that weren't enough, when she asks him if he's at least thinking about something, his only answer is, 'Nothing in particular.' This is not quite the same as if he had simply said 'No', but almost. When we are thinking of nothing in particular, nothing has any meaning for us anymore, rendering thinking itself practically redundant. This is absolutely unthinkable for a human brain that is constantly busy creating links between stimuli and behaviour concerning important relations and positive goals. It certainly annoys Herman's wife. She keeps nagging her husband until he finally loses his temper and shouts at the top of his voice, 'I'M NOT SHOUTING AT YOU!'

We do not know whether Herman's wife is just as unable to bear doing nothing and thinking nothing herself as she is unable to stand it with her husband. But we must assume that is the case. The life of *Homo sapiens* — who is, after all, not called *Homo inanis* (empty man) — is to a large extent ruled by the urge to banish emptiness or prevent it from occurring in the first place. In others, and in ourselves. And this characteristic appears to be particularly prevalent in today's affluent societies.

Smartphones over sex

There is hardly a moment when we do not have something to do, or at least something to consume. When we wake up, the radio is playing, then we check the news on our phones for the first time of the day as we partake of our breakfast cereal or toast. Then off we go to work, usually with the car radio playing or, if we travel by public transport, with our mobiles at full glow in our hands; some people even manage both. Once at work, we check our emails. And so it goes on, throughout the entire day.

Canadian researchers studied an average IT company in their country and found that employees were distracted from their work by incoming emails on average every five minutes, and interrupted their work to answer them within six seconds. In Germany, a survey revealed that 60 per cent of workers feel distracted by the flood of emails on their work computer. But no one makes any attempt to change this, for example, by putting themselves on an email fast. Under such conditions, it is a wonder that anyone ever gets any work done at all.

By lunchtime at the latest, our phones are out again — although your average worker can't hold a candle to school and university students in this respect. One survey at an American university revealed that male students spend around eight hours a day on their smartphones, and the figure for their female counterparts was ten hours a day. Sixty per cent of respondents said they couldn't exclude the possibility that they were addicted to their phones. Addicted or not, the fact that their phones are a hugely important part of young people's lives is shown by the results of a survey carried out by social-research institute Forsa. Six hundred Germans aged between 14 and 19 years were asked what they would be most able to do without for a period of one week: 70 per cent of the young women and 60 per cent of the young men said they would rather go without sex than their smartphone.

In front of the television in the evening, the older generation makes up for ground lost during the day when it comes to media consumption. According to Germany's Federal Statistical Office, over-50s in Germany spend almost 300 minutes a day watching TV. For the generation between 39 and 49, the figure is still approximately 220 minutes, which is still almost four hours.

Younger age groups watch less television, but when they do, they are often double screening by using the internet on their laptop or smartphone at the same time. As early as the year 2011, a survey carried out by Yahoo showed that 88 per cent of those under the age of 30 engage in this kind of media multitasking. It is not so unrealistic to extrapolate that figure today to almost 100 per cent — and not only in the USA.

However, before we succumb to the temptation to label media consumption as the main culprit for our inability to just do nothing, we should pause to remember that this behaviour is embedded in the nature of our society, which — just like other societies — forces its members into certain patterns of behaviour. And those patterns are to a large extent characterised by the fear of missing out.

Channel-hopper mentality

The term 'experience society' hit the scientific and political headlines in Germany in 1992 when the Bamberg-based sociologist Gerhard Schulze published his book of that title.[1] His thesis — which is well supported by empirical studies — is that people in the modern world see society's current orientation towards gathering experiences as an ideal. And those experiences are governed by a marketplace oversupplied with offers. Because the goal of that 'experience market' is to maximise profit, it will never be satisfied with what is already available — it must continuously create new offers. According to Schulze's analysis, 'while consuming a given experience, the urge to experience the next is already palpable'. Variety, says Schulze, is 'raised to the status of a principle', and the rate at which experiences are consumed becomes ever more frenzied.

An example of this from everyday life is our habit of channel hopping while consuming television output, which Schulze sees as a 'symptom of this general trend'. At the level of the brain, this phenomenon has its basis in the excitation circuit between the cerebrum and the basal ganglia, which prompts us to repeat any event which we have experienced as positive or rewarding in the past.

The problem with this general channel-hopper mentality in society is that, whatever we happen to be doing, we are already thinking of what we could be doing instead, or at least of what we might do next. According to Schulze, this robs experiences of their longevity, until they are barely able to provide more than 'instant, short-lived gratification', which leads to a 'permanent increase in appetite' for more experiences. This is in line with the thoughts of such philosophers as Epicurus and Schopenhauer, who saw as the main cause of humanity's 'vale of tears' the fact that those who wish to satisfy many wants can never become happy and free of desires, but at best happy and *full* of desires, because they will always be propelled from one desire to the next, with increasing voraciousness.

In the view of the Tübingen-based cultural theorist Hermann Bausinger, this is the point at which the 'experience society', in which the aim is to *do* something, becomes the 'results society', in which the aim is to *have done* it.[2] When experiences become so short-lived that nothing is really experienced at all, the question arises of what drives individuals to continue to crave them. Bausinger believes the answer is that the desire has now become not to *have* the experience, but 'simply to check it off a list'. Now it is only the result which is important.

According to a study carried out by SevenOne Media, the average channel hopper changes channel more than 140 times a day, and does not stick to one for even a hundred seconds. That is too short a time to absorb any significant content even from a morning chat show, let alone a feature film. No lasting experience and memory formation can take place in so short a time. It has been found that only two perceptions can take place. The first is, 'Ah, so that's what's on channel XY at the moment.'

The second is, 'How boring.' And then it's on to the next channel, up to 140 times a day.

Experiences are reduced to the staccato checking off of short-term results. And of course, this has consequences for our everyday lives beyond our TV-viewing behaviour. For example, it can manifest itself in inferring our level of popularity from the number of Facebook friends we have, in always feeling the need to work through at least ten different positions during sex, or in annoying restaurant staff with special requests. And when multi-billion-dollar, market-listed companies brag about double-digit growth in their worth, although that increase is due to nothing more than favourable currency exchange rates, or when lazy high-school students sue the education authority to force their school to let them graduate fully even without the necessary grades, then apparent results are seen as the goal, rather than achievement itself. Results are all that count, no matter how they were attained.

Eventually, the experience society, and its culmination in the results society, must also affect the minds of those who live in it. Not only in the fact that people begin to see their 'results quota' as confirmation of their existence. It is also interesting to observe what happens when the series of rapid-fire experiences and results is broken — and emptiness sets in.

Electric shocks over emptiness

When Stanley Milgram presented the results of his 'Behavioural Study of Obedience' in the 1960s, the world was shocked. The American psychologist persuaded ordinary, average citizens to operate a switch that, although they knew they were taking part in an experiment, they believed would deliver a shock of up to 450 volts to a person sitting in the next room. And they did this after they had already administered more than a dozen shocks of increasing strength, and the person in the neighbouring room no longer showed any sign of life. They did it without being forced or even threatened with consequences if they refused, and they had already received the payment they were promised for participating in the experiment. Not even one of Milgram's subjects aborted the test below a level of 300 volts; 65 per cent eventually delivered shocks of the highest available voltage to the person in the adjacent room, simply for failing to complete an absurd and impossible learning task.

Milgram's conclusion: virtually every human being has the capacity to become a willing tormentor if they should find themselves in a situation where torture is permitted. His results have now been qualified somewhat, not least because original film footage of his experiment shows that some of the subjects did put up considerable resistance. But, after Milgram's experiments, and in view of the bitter realities of human history, it cannot be disputed that people can generally be persuaded to torture or even kill.

Yet what circumstances could persuade them to torment *themselves* with electric shocks? Some 50 years after Milgram's study, Timothy Wilson of the University of Virginia decided to try to answer that question, and he set up a remarkable experiment to find out.[3]

The design of his experiment was less complicated

than Milgram's. Wilson and his team simply asked their approximately 400 test subjects to sit in an empty room and wait for 15 minutes. They were asked to spend the time thinking about a subject of their choice, but otherwise to do nothing. That meant not getting up from their seat and moving around the room. The uncomfortable furniture — just an uncushioned chair without armrests — made it impossible for the subjects simply to fall asleep. They had also been asked to hand in all their smartphones, iPods, binders of lecture notes, books, and anything else that might help them pass the time.

In interviews after their exile in the room, almost half of the test subjects categorised their time there as difficult to unbearable. Some nine out of ten subjects described experiencing mental unease during the test. They wanted to think intensely about something, as requested, but could not. Either they were unable to decide what to think about, or their thoughts wandered away from their chosen subject. 'They couldn't control the merry-go-round of their thoughts,' is how Wilson described it.

The American psychologist thought the unfamiliar laboratory situation may have made it difficult for the test subjects to relax, so he had them repeat the experiment in the familiar surroundings of their own homes. But, there, they had even more difficulty concentrating, and the lack of anything to do was even harder to bear than in the lab. The reason for this was that the test subjects were aware of all manner of possible distractions in their homes. And those distractions dominated their thoughts, making it impossible to concentrate on one topic.

When interviewed later, one-third of subjects admitted to having 'cheated' by getting up from their chair or even listening to music on their smartphone or iPod. There were no scientists supervising the subjects this time; they were asked simply to click on a link to a web program when they were alone and free of external distractions. I am sure the real number of 'cheaters' was much higher, as many of the subjects will have failed to reveal themselves as such in the subsequent interviews.

Wilson recognised that this empty lack of anything to do is almost unbearable for many people. They suffer. But how much do they suffer? Would these people go so far as to swap that situation for one that causes them suffering, but which at least offers a stimulus to occupy them? Wilson was put in mind of Milgram's experiments ...

For his next test, Wilson once again placed his subjects in a room for a quarter of an hour, but this time he offered them the chance of a little distraction. He installed a button that subjects could press to give themselves an electric shock. Not 400 volts, but a nine-volt shock, which does not cause pain as such but is rather unpleasant anyway. To exclude the possibility that his subjects were masochists with a liking for self-torment, Wilson gave subjects a trial shock before the experiment, then asked them how much they would be willing to pay *not* to repeat the experience. Most offered one or two dollars, and they were deemed fit to take part in the test. Those who offered less than a dollar or nothing at all were excluded from the experiment.

Wilson expected the vast majority of subjects would simply wait out their time doing nothing. After all, who would give themselves an electric shock with no prospect of a reward or compensation? But Wilson was wrong. Two-thirds of male subjects — about the same proportion as in Milgram's study — gave themselves at least one electric shock; the average number of self-administered shocks was just over seven per 15 minutes of doing nothing. And this figure excludes one particularly extreme case: one of the test subjects gave himself no less than 190 electric shocks in that time, although, at that rate, they can barely be called individual shocks, more a constant current.

Before we jump to conclusions and interpret this result as proof of the rampant smartphone and internet addiction of our times, we should note that Wilson's test subjects were aged between 18 and 77, so the sample included people who were unlikely to be spending much time on WhatsApp or Facebook.

What's more remarkable is the fact that only one in four women took advantage of the self-tormenting electrical boredom therapy. 'Women are less sensation-seeking than men,' explains Wilson; in general, men need more intense stimuli than women. They make up the overwhelming majority when it comes to illegal car racing on public roads, bungee jumping from bridges, binge drinking spirits, or munching on extra-spicy chilli peppers — and that's why they are so much more likely to give themselves electric shocks. But in general, women find emptiness just as unbearable as men; this was shown by the results of Wilson's original experiment. Other experiments even indicate that women in such situations are quicker to lose patience and become aggressive.

We can conclude that the feeling of emptiness is almost unbearable for most people. Given the alternative, most men will opt for a painful stimulus over no stimulus at all. And the more people believe that there is no alternative to emptiness, the more unbearable they find it. This fact in particular is important to keep in mind because it will play a central role later in this book, when we consider emptiness not as a dreadful fate, but as a way of achieving relief for our overworked minds.

The brain wants effects

Before we get to that, we should first examine why it is that emptiness is so unbearable for us in our day-to-day lives. Why couldn't the subjects in Wilson's study just indulge in their thoughts for 15 minutes? What is so terrible about disconnecting from the daily grind for a while and letting our brains run idle without any external stimuli? There are no end of relaxation courses, meditation manuals, and advice books that recommend exactly that; they advise simply doing *nothing* to find inner peace. And, given the opportunity to do that as part of a scientific study, people turn out to be incapable of it. But why?

It seems clear that the 'multi-option society' we live in plays an important part in this, as it constantly provides us with some means to keep us occupied. This is especially thanks to social media. These platforms bombard us with a bright, shiny array of easily accessible ways to occupy ourselves — all we have to do is switch on our smartphones or any other web-enabled device! — to such an extent that we practically go 'cold turkey' when we are cut off from that source of distraction.

However, people were already trying to escape boredom and emptiness long before such options for distraction and diversion existed. Even the ancient Greek philosopher Diogenes railed against the hectic, hustle-and-bustle lives of his fellow Greeks, and, in the 17th century, the French thinker Blaise Pascal complained that people are unable to sit happily and quietly in their own chamber, describing how they become ill-tempered without 'what is called diversion'.

In the 1950s, the Canadian neuropsychologist Donald Hebb paid his test subjects 20 dollars for every day they spent subject to extreme sensory deprivation (this is described in more detail in Chapter 6). Hebb's test subjects were cut off much more radically from the outside world than those who took part in Wilson's experiment. Nonetheless, the paid volunteers expected they would easily be able to sit out the time and take the money. Yet most subjects abandoned the experiment after just two days, and not even one managed to stay in the sensory-deprivation conditions for an entire week.

Remember: this experiment was carried out in the era before smartphones, when people had to make do with black-and-white television and the radio. This shows that the phenomenon is nothing new and that this inability to endure emptiness and inactivity is not just a sign of our times. Rather, it is a typical feature of our brain. The brain is a driven organ, and the driving force is the brain itself. More precisely, it is the brain's mesolimbic dopamine system and a few other areas of the brain, which are often called its 'positive reward centre' and which we have already met in the guise of the excitation circuit between the cerebrum and the basal ganglia.

This reward system is made up of cells with long projections, which begin at the interface between the midbrain and the interbrain (or diencephalon) and penetrate deep into the frontal regions. There, depending on the how rewarding the stimulus is, an amount of the neurotransmitter dopamine is released, which triggers the urge to achieve an effect. Now the brain is galvanised into activity aimed at attaining its goal. The dopamine system continues to fire, especially if the expected effect doesn't materialise; the brain wants to correct that undesirable state of affairs.

The brain can use certain mechanisms that compare the expected and actual effect of its actions to ascertain which of those actions were useful (i.e. which resulted in the greater reward). But what the actual content of those actions should be depends on what the brain has learned previously in this context. As an extremely plastic organ, which almost seems to have been designed to be shaped and reshaped, the brain is basically uninterested in what goes on in the outside world.

This neutral lack of interest is only superseded when the brain learns what is important to itself. This can take on an egotistical or altruistic character. It can be loving and gentle or brutal and bullying in nature. It may be associated with the survival of the species and reproduction, or it can be associated with pain and destruction. Some people sacrifice their lives for others, and some people sacrifice others' lives for themselves. And under certain circumstances, it can make sense to give yourself an electric shock. When it comes down to it, it's all one and the same to the brain: just one among many possible goals towards which it can orient its interests and desires.

In summary: the brain 'wants' effects. These are preferably effects that the brain has already evaluated as being positive, or which have counteracted a negative effect in the past. However, the yardsticks for measuring this become lost when the only alternative is emptiness. That's when the brain tells the hand to press the button that triggers the electric shock, because even that is better than striving for *no* effect. If we can even speak of the basic nature or character of an organ that is so extremely flexible, then its nature is principally founded on its desire to achieve effects — and by *their* very nature, effects are the precise opposite of emptiness.

So we should not attach too much importance to the influence of the results society we live in. With its multi-optional media landscape, it may indeed contribute to our inability to bear boredom or emptiness. But the only reason it is able to do so is because it is so geared towards satisfying our brain's craving for effects. We perceive emptiness as unbearable, even threatening, because the results society and our effect-hungry brains mesh so well. The construct of modern society has not conquered our central nervous organ, but rather provides it with an overabundance of precisely that which it craves: effects.

And when effects don't materialise, the brain reacts with confusion and panic, and lashes out.

Quite apart from this, there is also a simple evolutionary and historical explanation for the fact that our results society and our brains fit together so well. Societies do not come from nowhere. They are ultimately the product of human brains — and why should those brains create something that does not correspond to their own nature? Milgram showed that brutal political systems such as fascism can develop anywhere and in any society because people's brains have been instilled with a strong tendency towards control and obedience. In the same way, the multi-optional, results-based society is not something that has been imposed on us unwittingly. It, too, was thought up by the brains of its members.

The brain can also 'do' emptiness

It seems doubtful that any ethics commission would give Hebb permission to carry out his experiment today, since it involved exposing his test subjects to one of the worst forms of torture possible: that of emptiness and isolation. Yet, as we will see, this experience is to a great extent due to the fact that we approach emptiness with the wrong mindset, seeing it as something to be feared, a loss, and, as such, something to be resisted. In fact, emptiness need not be a calamity.

This is because our brains do not only want to achieve effects — that is just one aspect of their vast repertoire, and one that is particularly served by our results society. There is another side to the brain, where effects and functions play no part, and are even expressly blanked out. In recent times, we have lost sight of this somewhat, but this does not mean it is gone forever. Our brains *can* deal with emptiness, and this ability is as much part of the brain's nature as the desire for effects. And, of all people, it was philosophers — that is, people whose heads are particularly full of thoughts — who were the first to draw attention to this.

2

Free at Last

philosophers as pioneers of emptiness

'Man still prefers to want nothingness, rather than not to want at all.' This is how Friedrich Nietzsche summed up what modern science might describe as the prime feature of the human brain, and its greatest problem at the same time. That is, that it somehow always *wants*, that it is oriented towards achieving effects. And there is no hope of this craving becoming any less intense as we grow older, although the organs of the brain responsible for its executive functions do lose freshness and power, making it increasingly difficult for the brain to get what it wants. With time, habituation effects also reduce our sense of satisfaction at achieving a desired goal. But neither of these effects does anything to lessen the brain's will. In fact, the opposite is true.

The field of motivational research has taught us that, although there is little change in the things we like over time, the will to gain them becomes ever stronger. This is because the attractiveness of the positive stimuli associated with that attainment increases *every time* we experience that satisfaction, irrespective of the fact that our subjective feeling of satisfaction grows less intense over time. This reduction in intensity even serves as an additional incentive and spurs on the will. It's as if the brain were saying to itself, 'Okay, that wasn't very satisfying this time, but next time is bound to be better ...'

Virtually every kind of addiction can be explained by this mechanism. Heroin addicts, for example, often describe how their craving for the drug continues to get stronger, although they no longer gain any great satisfaction from a fix. Even the infamous phenomenon of the 'lecherous old man' can be explained by this mechanism. The fact that old men continue to pursue young women even when their physical prowess between the sheets now leaves a lot to be desired shows that our will gets stronger, rather than weaker, with time. And the same is true of almost everything we feel drawn to. We want those things more and more, although they offer us less and less.

This may sound absurd, but it is the way our brain works. It prefers to want nothingness than not to want at all. And on closer inspection, this no longer appears quite so absurd. It makes perfect sense from an evolutionary perspective and the law of 'survival of the fittest': when a living being senses its increasing frailty and tries to compensate by trying harder to exert its will, it might squeeze a few more revs out of its sputtering motor and increase the creature's chances of survival.

Most philosophers find this constant striving for effects unacceptable. If we are always propelled towards the object of our desires by our will, we cannot consider ourselves free. And if the will becomes stronger and stronger with time, then it must follow that old age does not mean we cast off the tribulations of life as we get older, but that we increasingly become slaves to them. Thus, there is no hope of salvation, and we can only expect to become less and less free as we age. This makes a mockery of books with titles such as 'The Freedom of Old Age' or 'The Happiness of Old Age'. Those philosophers who see themselves as the guardians of freedom especially will find this impossible to accept. Some have come up with concepts for neutralising the brain's craving for effects and therefore freeing us from the dictatorship of the will. And one of those concepts — probably the most important one — is the doctrine of emptiness.

Heraclitus's question: What remains if everything is in flux?

If we take an unblinkered look at the concept of emptiness, we can state that there is one thing that it is definitely not: substance; or material, if that's the word you prefer. Everything that defines our lives is made up of material, and so Western philosophers initially saw no reason to doubt it.

In the sixth century BC, Greek philosopher Anaximander ascribed attributes such as indivisibility and boundlessness to his *apeiron*, while considering it the 'primordial substance', something profoundly material in nature, out of which everything that exists is born. And Pythagoras, open though he was to all sorts of metaphysical larks, was unable to imagine a world beyond the material; he even imagined his beloved numbers to be tiny objects of a certain thickness. Then along came Heraclitus of Ephesus and turned everything on its head.

Heraclitus differed from his fellow philosophers even in his self-image. He was proud of the fact that he never followed a teacher and never had any students. His belief was that 'much learning does not teach understanding'. Rather, for him, the absence of any preconditions was the primary precondition for the formation of thoughts that do more than just replicate the blowhard blustering of their own time, and really approach truth (or the *logos*, as he termed it). One could describe Heraclitus's guiding principle as 'emptiness rather than education'. This freed him from the fetters of material thought.

The most famous aphorism to emerge from his writings is often misunderstood: *panta rhei* — everything flows. According to this, everything is in a constant state of flux, subject to never-ending change. 'No man ever steps in the same river twice,' Heraclitus explains. Not only is the water no longer the same as the first time he stepped in the river, the man himself is also no longer the same.

What most people fail to realise about *panta rhei* is that emptiness features greatly in the idea. If things are in a constant state of change, it follows that they have already ceased to exist as they once did by the time we perceive them. It is striking how apposite this realisation is to the workings of the brain. As we go about our daily lives, our brains are constantly required to react — to visual and acoustic stimuli or to signals from our organ of balance and receptors in our muscles and other parts of our bodies. However, most of these changes are short-term and negligible and do not leave any lasting traces: a couple of firing neurons, a couple of electrical charges — and then it's over.

For something to entrench itself in our brain, it needs to have significance. This stabilises the relevant pathways in our neuronal network. But even that lasts only until it is overwritten. By something that — for whatever reason — is considered more important by the brain. Then the first memory is forgotten and the new one is stored, until it in turn is overwritten and forgotten. Both in the short-term and in the long-term, nothing remains the same in the brain; it is in a state of constant change — which brings us back to the banks of Heraclitus's river.

When Heraclitus dipped his toe into the river, he did not know about the fleeting electrical currents flowing inside his skull. Nevertheless, he reached the conclusion that the world as we perceive it contains its own demise. It is, and then it's gone. Emptiness and nothingness follow hard on the heels of existence. For this reason, Heraclitus refused to attach any importance to things, fleeting as they are. The world? Nothing but 'a heap of rubbish piled up at random'. Men? 'Many are bad and few good.'

According to legend, Heraclitus eventually retreated into solitude, losing not only all contact to his fellow human beings, but also his ability to use language in the normal way. Thus he was dubbed Heraclitus the Obscure, since his utterances were so difficult to understand. He did write one treatise, called *On Nature*, but barely anyone was able to follow it. Not even Aristotle; and Socrates said of it: 'The part I understand is excellent, and so too is, I daresay, the part I do not understand; but it needs a Delian diver to get to the bottom of it.' The Delian divers were famous in the ancient world for their ability to descend to depths where darkness is almost total.

For example, Heraclitus's treatise contains formulations such as 'Eternity is a child playing, playing checkers; the kingdom belongs to a child.' Or, 'The bow is called life, but its work is death.' Before we start trying to make sense of these sayings, we should remember that Heraclitus may not have intended them to be understood at all. His aim might have been to seek a sense of emptiness. A *sense* of emptiness, mind you, not an *understanding*, since nothingness cannot be explained.

Heraclitus liked flirting with the inexpressibility of emptiness and enjoyed being stared at open-mouthed, or even abandoned in ire by his fellow human beings because of it. His follower Cratylus took this one step further and eventually gave up speaking altogether, since language can never describe the origin of the world and therefore also its nature. Cratylus would simply raise a weary finger when anyone asked him a question.

These might sound like nothing more than the bizarre escapades of eccentric philosophers, but modern brain research has shown that the language areas of the cortex actually do remain inactive when the brain's owner is reacting to emptiness. Furthermore, Cratylus's weary finger lifting seems almost like entertaining small talk compared to the *kōan* of Zen Buddhism.

The principle of Buddha: thy will be extinguished!

Gautama Buddha lived at about the same time as Heraclitus, and he had a similar view of existence: that it is a sequence of fleeting moments, which disappear even as we perceive them. According to this view, the universe is a ceaseless stream of individual moments of being, or, as a follower later expressed it, 'a continuum of transience'.

Where Heraclitus saw philosophy as a product of introspection and mainly turned his attention to nature, Gautama went one step further, or, more accurately, one step back, applying the principle of emptiness to our inner life. Ultimately, this principle means there can be no persistent self. Consciousness is also seen as constantly renewing itself, at every moment. It is only the speed of our mental processes and their entangled nature that create the misleading impression that there is a permanent self at their source.

We think or feel something at a given moment and everything merges together so quickly and seamlessly that we have the impression of sitting on a train that carries us through the world. And, since nothing exists except our thoughts and feelings, this train must be our 'self'. But in fact, according to Gautama, neither the train nor the world exist. There are only moments, which light up for an instant and are gone again. They appear as feelings and thoughts, but there is no one leading them, and certainly no 'self'.

For Gautama, then, any consideration of the past is pointless, since, if the course of time has no context but consists only of a sequence of individual moments, then there can be no meaningful history. And this is why, just like Heraclitus, the ancient Indian teacher attached no importance to historical or philosophical traditions.

Far more important for Gautama was the question of whether it is even possible to exist in this illusory edifice of being, in which we imagine ourselves to be a persistent 'I', marching through time and the world. He concludes that it *is* possible, because: why should it not be possible to live in an illusory edifice? It is, in fact, what most people do.

But this means we are actually living inside a wheel of eternal life. First, because, as inhabitants of an illusory edifice, we live in a state of permanent deception and therefore are bound to be disappointed again and again when we realise that things do not always turn out as we imagined and, most importantly, when we realise that things are not as permanent as we had hoped.

Second, because the idea that people and things are connected in a meaningful way drives us to *want* something of them. We expect things, animals, and people to make us happy somehow, and this leads us to put pressure on ourselves — despite the fact that, in a world without permanence or substance, there can never be any lasting satisfaction, and therefore no enduring happiness. Or, to put it another way, we are constantly snatching at happiness, but we will never catch it, since that is impossible in a world of meaningless change. All we gain from this vain striving is suffering.

'Birth is suffering, ageing is suffering, sickness is suffering, separation from the loved is suffering, association with the unbeloved is suffering, not to get what one wants is suffering,' declares Gautama, 'all life is suffering.'

It is remarkable that Gautama could pinpoint one of the characteristic features of our brain as long ago as 2,500 years. This feature is the brain's constant wanting, and the fact that although such desire can never be completely satisfied, the brain repeatedly fuels its own need by achieving partial goals. Yet Gautama was not alone in thinking this way. Even before his time, the collection of ancient Indian writings known as the Upanishads expressed similar ideas when they branded the 'thirst for life' as the source of all evil. And the Ancient Greeks' legend of Pandora, who unleashed all the evils of humanity when she opened her infamous box, leaving only hope inside, was also a way of explaining how human beings will always cling hopefully on to life and face new torments, no matter how terribly they have already suffered. What was new in Gautama's teaching was the solution he suggested to escape this bleak cycle of suffering.

His suggestion is to extinguish the will like a campfire starved of wood. So, not — as was common in his time — by means of ascetic exercises that confront the will, only to make it appear more intensely by turning it into an opponent. And also not by means of drugs or sleep, which merely numb the will or suppress it temporarily, but cannot extinguish it permanently. Rather, Gautama advises us to attempt the impossible and base our existence on something that isn't in fact anything — that is, nothing. In real terms, this means we should give up our frantic attempts to fill our lives with content or meaning, but open ourselves up unconditionally to the emptiness of existence — and abandon the idea that we are an effective self wandering through space and time. Such a practice, Gautama continues, will lead to an extinguishing of the will — upon which we enter the state of nirvana, liberated from the cycle of suffering.

We may feel that we are departing here from the familiar paths of thought that we normally follow through life. The self as a vehicle of travel through space and time disappears, and instead we empty ourselves of everything, including our will, and enter nirvana. This idea is often met with rejection, ridicule, or even fear. And Gautama did not exactly do much to explain, because he did not see himself as a saviour but rather as an inspirer, whose task it was to nudge people gently in the right direction but otherwise to let them make their own way.

In Zen Buddhism, however, this emptiness takes on concrete forms. So concrete, in fact, that it can barely be voiced.

The silence of the Zen masters: emptiness leaves us speechless

It is told that immediately after his birth, Buddha pointed with one hand towards Heaven and with the other towards the Earth, then walked seven steps in a circle, glanced in each of the four directions of the compass and eventually proclaimed:

'In Heaven and on Earth, I alone am the Honoured One.'

Of which Master Yunmen said, 'If I were a witness to this scene, I would have knocked him to death with a single stroke and given his flesh to the dogs for food — this would have been a noble contribution to the peace and harmony of the world.'

This is a rather violent tale. But it is not from an opponent of Buddhism, as you might think. In fact, it is taken from the book *From the Record of the Chan Master 'Gate of the Clouds'*, one of the central works of Zen Buddhism, written by Yunmen, who lived from 864 to 949.[1] He is also famous for replying to the question 'What is Buddha?' with the answer 'A dry shit-stick.' Such jokes are typical of the attitude of this philosophical movement. They are a negation of the role of Gautama Buddha as a pioneer leader and this finds expression in other sayings of Zen, such as 'If you meet Buddha on the road, kill him.' But despite such sayings, doubt is never cast on Buddha's central teaching — that liberation from the wheel of suffering is to be found in nirvana. On the contrary. Such aphorisms take this teaching to its logical conclusion — to such an extent that the personality of the Buddha is completely negated.

Zen (from the Chinese, *chan* = state of contemplation) originated in China in the fifth century AD, but assumed its decisive form seven centuries later in Japan. It is much more 'ascetic' in its use of language than any other form of Buddhism, because it rejects itself as a doctrine. According to its creed, words are a completely inadequate vehicle to communicate the nature of Zen. Rather, Zen communicates itself independently of words and writing by 'looking into its own nature and becoming Buddha', as the Zen masters like to put it. For this reason, their philosophy does not exist as a system of thought, but as an unorganised collection of instructions, aphorisms, and stories, which often seem funny, almost always paradoxical, and definitely opaque to those of us from Western cultures.

A typical example is the tales of the ninth-century Zen master Daian. He once said 'Being and non-being are like the wisteria winding around the tree.' A Zen disciple named Sozan heard of this and undertook a long journey in order to put his question personally to the master. 'When the tree is suddenly broken down and the wisteria withers, what happens?' That's not such a silly question, since it addresses the problem of how we are ever to get over the contradiction of being and non-being. But Master Daian was engaged in building a mud wall. Upon hearing the question, he upturned his wheelbarrow and, laughing loudly, walked away. In his frustration, Sozan put his question to another master, who answered in a similar way. But this time Sozan understood. He smiled, bowed reverently, and walked away.[2]

We could now set out to analyse this story. To speak of how we will never recognise even a speck of truth while we are filled with ideas of being and non-being and an urge to explain them. But this would not be true to the spirit of Zen. The Zen masters would call upon us to study the story, but not by reflecting on it in search of a higher meaning within it. That would be nothing more than juggling with concepts, without reaching any conclusions, and therefore without reaching enlightenment and nirvana. Instead, the Zen masters would tell us, we should meditate on the story. But what is the difference between meditating and thinking?

The best way to answer this question is to compare Zen meditation with the kind of meditation practised by René Descartes in the 17th century. After much musing in his bed, the French philosopher reached the conclusion that all can be doubted but one thing, and that is that someone exists to do the doubting and the musing. Amid all the doubt, one certainty endures — that of the existence of the self. Now let us imagine that Dōgen, one of the most influential teachers of Japanese Zen Buddhism, enters Descartes' chamber, sits down on his bed, places a fatherly hand on the philosopher's slender shoulders, and says, 'That's right. But now descend deeper into your meditation and take your doubt further, until you yourself become the great doubt in which the self eventually breaks down.' Descartes would probably have banished the master from his chamber in indignation. But if he didn't, and — following Dōgen's advice — took his doubts further, he might well end up shouting out for joy, 'Neque cogito, neque sum — I neither think, nor am!'[3]

The philosophical meditation practised by the rationalists was aimed purely at reaching ultimate certainties and thus gaining control over life. Practitioners of Zen meditation, on the other hand, leave all certainties behind them, ultimately gaining release by attaining nirvana. Rationalists are driven to seek solutions and certainties, Zen practitioners, by contrast, extinguish that drive by descending into emptiness.

Later, we will see that the activities observed in the brains of practitioners while they are meditating are different from those we see in the brains of people who are thinking or sleeping. It is interesting to note that, long before the invention of modern measuring technology, Zen Buddhism recognised the fact that while human beings are principally oriented towards control, activity, and effect, they also have the ability to shift to 'empty mode'. This does not say anything about whether that makes human beings happier, as Western followers of Buddhism like to claim; the ancient Zen masters considered happiness to be just as illusory as the self. But the fact remains that emptiness is different to the thinking or acting we engage in throughout our day-to-day lives. And Zen Buddhism must be given credit for opening a window on emptiness for us.

Schopenhauer and music: note by note into emptiness

In his magnum opus, *The World as Will and Representation*, Arthur Schopenhauer manages the trick of combining Eastern Buddhism with Western rationalism.

Very much in the spirit of Buddha, he begins by declaring that life is constant suffering and the cause of that suffering is the will, which works from 'deep down' in our unconscious mind. It is the reason life swings into action at all, the reason we eat and drink, desire each other, and reproduce. It determines whether and how we act and think. The will is always driven to strive for fulfilment, and, as long as that fulfilment is not achieved, we suffer pain — yet when it is achieved, we are assailed by boredom. Pain and tedium are what determine our existence, which is why Schopenhauer believes, 'Every life story is a story of suffering.' Bleak.

And the epistemological side of life appears no less bleak in the philosophy of Schopenhauer. His views proceed from the ideas of Immanuel Kant, who tells us we live in an illusory world and are far from getting to the bottom of his famous 'thing-in-itself'. Or, in Schopenhauer's words: '[Humanity] knows no sun and no earth, but only an eye that sees the sun and a hand that feels the earth.' Thus, for Schopenhauer, in contrast to Buddha, the main problem is not the mutability of the world, but the fact that it can never be perceived. Both eventually come to the same conclusion: that the world is essentially characterised by suffering and deception.

So how can we escape this world, which Schopenhauer liked to describe as a 'vale of tears'? Asceticism and sleep are not options for the Western philosopher, and suicide is also out of the question, since it is a product of desperation and thus of an overheated will and as such cannot be a release from the will. Schopenhauer believes a suicidal person has a will to live, and is only dissatisfied with the conditions under which life has presented itself to him or her. This is why he believes suicidal people by no means surrender the will to live, but only life, in that they destroy the 'individual manifestation'.

According to Schopenhauer, we should follow two philosophical paths to escape the vale of tears. One is ethical in nature, calling for renunciation and sympathy

— renunciation as a way of breaking the power of the will, and sympathy as way of overcoming our own subjective suffering through knowledge of the suffering of others. In this, Schopenhauer once again approaches the ethical stance we also find in Buddhism.

However, with the other path, that of aesthetics, Schopenhauer breaks new philosophical ground. His concept is that art and music allow us to see beyond the phenomena itself. That alone would be enough to release us from our illusory edifice. In the case of music, there is a second significant aspect: the fact that it is a direct manifestation of the thing-in-itself. Or, to put it another way, music enables us to feel what things would feel themselves, if they had the capacity to do so. And what we feel, says Schopenhauer, is nothing other than the will. Music is the echo of the will that is at work in things. It is not a depiction of things from the outside, but a realisation of their own will, which resides within them.

Take, for example, a flower just about to bloom, and a skilled pianist who expresses this scenario in music at the keyboard. When the musician plays, he renders the flower's will to blossom immediate and alive in himself, and also in us as listeners. When we hear this musically composed flower, we experience it exactly as if we were the flower ourselves. Meanwhile, our personal will is extinguished; we *empty* ourselves; our own moods, sensitivities, needs, and intentions retreat — and eventually we feel only the will of the flower. Some people experience this effect particularly keenly in the lieder of Franz Schubert.

Among empirical scientists, there will certainly be a

sharp intake of breath here, as this is where Schopenhauer leaves the solid, facts-based field of natural science and enters the speculative, playful realm of metaphysics. On the other hand, he also repeatedly stresses the fact that acts of will and acts of the body are not two different things linked by their causation, like a sort of 'ghost in the machine' causing the initial spark for our finger to begin to move. Rather, he says, the two are the same thing, distinguished 'only through the form of knowledge into which they have passed'. The actions of the body are merely the objectification — in the sense of 'making visible'— of acts of will.

For Schopenhauer, the will is a construction not of the mind, but of the body. And this is why feeling the will of a flower in music is definitely 'concrete'. This is also in agreement with the latest findings of brain research, which indicate that the areas of the brain involved in spatial orientation, such as the hippocampus and the superior parietal lobule in the upper part of the parietal cortex, have a clear tendency to blur the border between the self and the outside world under certain conditions — and music is one of the things that can create those conditions. We will return to this in more detail later.

Furthermore, Schopenhauer had no problem with the natural sciences. He had read the works of the French physiologists of his time and was very well acquainted with their theory of the brain as the organ of thought: 'Knowledge in general is known to us only as a phenomenon of the brain, and we are not only unjustified in conceiving it otherwise, but also incapable of doing so.' For Schopenhauer, the idea that nothing divine or otherwise transcendental will necessarily come from this trivial, corporeal basis was not a problem: 'The brain thinks, just as the stomach digests.' As the latter organ prepares food for its passage through the gut, the brain processes perceptions of our world so that we can digest them. Such parallels are not of the kind drawn by someone sitting in an ivory tower wanting to protect the noble world of thoughts from being dragged down to the base level of the corporeal.

Simply crazy: Nietzsche and the intoxication of emptiness

Schopenhauer's idea of release from the will was similar to that of Gautama Buddha — like the gentle extinguishing of a fire, silent and unspectacular. This also explains his great aversion to anything that loudly and insistently draws attention to itself. Political uproar and great celebrations, captious speakers and vain self-promoters, lurid or romantic novels, garrulous professors such as Hegel, or bombastic composers such as Wagner — they were all abhorrent to him. He preferred a stroll round the gardens with his pet poodle or the isolation of his chamber, playing his recorder. Even harmless sociability was among the 'dangerous inclinations' Schopenhauer decried, springing as it did from people's 'inability to endure solitude, and thus themselves'.

Those words could just as easily have come from Friedrich Nietzsche, who was also a fan of silence. That could have been due to the constant bouts of headaches he suffered, which forced him to spend a total of seven summers in seclusion in the Swiss Alpine village of Sils Maria. Still, that more than anywhere else was the place where he was able to immerse himself in emptiness:

'I sat here, biding, biding — but for nought,
Beyond good and evil, now the light
To savour, now the shade, all merely mime,
All lake, all midday, all untending time.'

Although Nietzsche was not yet 40 years old when he wrote those lines, he had already reached the autumn of his philosophical career. He had left Schopenhauer and Wagner — for whom he had initially been inflamed with adoring admiration — behind him, and finally developed into the 'destroyer of all morals'. He dismissed all the achievements of the civilised world, such as morality, Christianity, middle-class education, and science as the unnatural figments 'of declining, weakened, weary, condemned life'. He considered all of these as having failed, especially Christianity, since, as he infamously put it, 'God is dead.'

In the seclusion of the Swiss Alps, Nietzsche then wrote his book *Thus Spake Zarathustra*, in which he formulated the counter-concept: the Dionysian, named after Dionysus, the Greek god of intoxication. This was, in part, a call for people to overcome all that is reasonable and instead to heed the 'voice of the body': to reject the comfort of a world that is morally protected, thoroughly explained by science, and comforted by religion, in favour of a life of risk, in which nothing is predetermined and which is completely empty of content. There are instincts and spontaneous intuitions, but no thinking guided by any kind of self or aimed at any particular goal. However, there is also no inevitable silence, which is where Nietzsche differs from Schopenhauer and Buddha.

This slight, highly sensitive, sickly, desperately noise-avoiding thinker of Sils Maria interprets emptiness not in terms of the withdrawn, unworldly monk, silently sitting in the lotus position for hours on end. Nor does he see emptiness like the recorder-playing Schopenhauer, with a kind of musical devotion in mind. Nietzsche viewed emptiness far more in terms of ancient Greek Dionysian festivals, where people danced themselves into a frenzy, caterwauling and carolling to succumb to the 'collapse of the *principium individuationis* (the principle of individuation)' and reach 'the highest intensity of all [their] symbolic capabilities'. Nietzsche's empty people are in exuberant motion, communicating loudly and without inhibition, albeit also without words.

In Chapter 8, when we examine the topics of sex and epilepsy, we will see that emptiness does not necessarily descend on people in the form of a silent moment. In locked-in and dementia patients, as well as Zen Buddhists, emptiness is silent and motionless, but orgasms or ecstatic dancing are precisely the opposite. It is in the nature of emptiness that it cannot be pinned down there either — and it is to Nietzsche's credit that he communicated this fact to us 'with a hammer', as he put it. This also fits in with his reference to the 'eagle courage' we need to 'slay the giddiness' that overcomes us when we enter a state of emptiness. In his *Zarathustra*, Nietzsche has a tightrope walker plunge to his death when a devilish jester suddenly jumps over his head on the rope. His warning message is that such a fate could befall anyone who leaves everything behind them and commits to nothingness. But the reward is the 'delight in destruction' we feel.

Nietzsche was to experience that feeling at first hand. In 1889, the emptiness he so extolled overcame him in very real way, in the form of a mental and physical breakdown. At first, he began writing letters saying he intended to imprison the Pope and have Kaiser Wilhelm shot — and chose to sign those missives 'Dionysus' or 'the Crucified One'. He also began dancing naked in his Turin hotel room and requested that all the pictures be removed from the walls so that it would more closely resemble a temple. He would play the piano manically by night, and by day he would hold forth to himself about being the successor to the 'dead God'.

Eventually, his hotel landlord was called to fetch him from a local police station after he caused a public nuisance on the streets. The legend that tells of him rushing to hug a donkey that had been whipped by its owner is unlikely to be true, since Nietzsche was not exactly known as an animal lover. Whether that story is true or not, he had lost his grip on reality, and normal conversation with him was all but impossible. His friend Franz Overbeck wrote, 'He, the unparalleled master of expression, was unable to articulate even the delights of his gaiety other than with the most trivial of utterances or through bizarre dancing and leaping.'[4] Nietzsche had apparently gone from being a theoretical 'Dionysian' to a real one.

Even while Nietzsche was still in Turin, a German medic called Dr Baumann made a devastating diagnosis: insanity. He had seen Nietzsche for barely a couple of minutes, and heard how his patient was constantly asking for food and 'women's rooms'. Despite the fact that Dr Baumann's diagnosis was more reminiscent of a snap judgement than an expert consideration, it was barely questioned from that point on. The neurologist and psychiatrist Wilhelm Lange-Eichbaum went even further, describing Nietzsche as 'so demented' that he might be compared to a 'completely burnt-out crater'.[5]

Lange-Eichbaum was also the originator of the explanation for the philosopher's mental illness that is still generally accepted today: syphilis. He explained that the infection had gradually destroyed Nietzsche's brain and caused his insanity. This diagnosis has now been challenged, not least of all because Nietzsche probably never had sex in his entire life, and syphilis is a sexually transmitted infection. Research carried out by Leonard Sax of the Montgomery Centre in Maryland revealed that Nietzsche did not display many of the symptoms typical of syphilis.[6] For example, he did not have the uncontrollable trembling of the tongue that is a typical sign of the disease, explains Sax, whose research covers both psychology and the history of medicine. Nor did the patient present with an expressionless face and slurred speech; he was still able to speak and write long after his breakdown in Turin. The only abnormal physical finding was an asymmetry in the size of his pupils, but even that cannot be taken as a sign of neurosyphilis, since, as Sax points out, 'Nietzsche's right pupil had been larger than the left since early childhood.'

The fact that the philosopher lived for another 11 years after his breakdown in 1889 should also give pause for thought, as this is far longer than would have been possible for a syphilis patient in those pre-antibiotic days. Most syphilitics died within five years of their first symptoms appearing. Indeed, Nietzsche's doctors initially gave him just two years to live. For Sax, the conclusion is clear: 'When examined closely, every aspect of the syphilis hypothesis fails.' So we simply do not know what was going on inside Nietzsche's brain.

In any case, Nietzsche cannot really be said to have suffered. In the first few years of his derangement, the philosopher went on long walks, often improvised at the piano, and applauded kindly whenever someone else played. His friends had to accept the fact that he no longer recognised them when they came to visit him, but they affirmed that he 'did not look bad', was 'dreamy' and 'tranquil', no longer seemed 'as if he were sick at all', and had 'something natural' about him. This man who before could get upset about almost anyone and anything, only ever had one more temper tantrum — when the writer and anti-Semite Julius Langbehn tried to convert him to Christianity. Otherwise, Nietzsche was, as his mother liked to put it, 'a child of the heart'. He later became increasingly lethargic and spent much of the time in bed staring into the void, but he certainly did not appear to be unhappy or suffering.

Nonetheless, Nietzsche's condition in the final years

of his life can, of course, be described as tragic, in view of his former great eloquence. In the same way, the fact that his sister eventually turned him into an exhibit — dressed in an angelic white cloth — and bastardised his philosophical legacy appears humiliating and inhuman. Yet all this no longer affected the patient himself. Through emptiness, he had found a kind of peace that had mostly eluded him during his creative years. The brain of the philosopher had reached a state that — viewed from the inside, from the point of view of the man himself — was neither catastrophic nor tragic, but rather a blessing.

No sleep and no hope: Emil Cioran and the ecstasy of capitulation

The Romanian philosopher Emil Cioran picked up the baton of emptiness from Buddha, Schopenhauer, and Nietzsche and combined it with his childhood experiences in the Carpathian Mountains in the early 20th century. Living conditions were primitive, even pre-civilised, and no one tried to force the people there into any kind of system or to exploit them for any purpose. This only reinforced Cioran's later opinion of civilised life, including the study of philosophy, as 'enslavement' and 'despotism'. As a young philosopher, he rebelled against everyone and everything, and his anger grew so great that it caused severe insomnia — seven years in which he felt he almost never slept!

However, it was in Cioran's nature that he was able to utilise this condition for his kind of philosophising. He experienced those 'white nights' as a rebellion against bourgeois life, with its Calvinist ethic, in which senseless working alternates with senseless sleep. And he saw his insomnia as a revolt against rational thinking and therefore also against conventional philosophy: 'There is … no system which resists those vigils. The analyses of insomnia undo all certainties' — and lead to new realisations:

'During sleepless nights we follow the course of time backwards and experience once again primeval fears and pleasures; events which took place long before our history, long before our memories began. Insomnia causes a return to the origins, transports us to the dawn of humanity. It banishes us from temporality and forces us to listen to our very last memories, which are also our very first … Insomnia, in a single night grants more knowledge than a year spent in repose.'

It is at this point, if not before, that Cioran begins to seem strange to us. If we have a sleepless night every now and then, we feel exhausted and far from having learned anything. Also, we know that it is during sleep that our memories are consolidated and stabilised, which is why more and more employers are providing nap rooms for their workers in the hope that they will be more efficient after a little sleep. So how can insomnia provide us with new ideas?

Perhaps we should add one aspect to the words of the philosopher to make it clear what he meant. When we speak of *other* or *detached* ideas, it becomes clear that what we are talking about is something quite other than sleep. Sleep regenerates us, making us fit again to face our waking hours. Cioran is not at all interested in that, because it does nothing to liberate us from the hamster wheel of work and recharging for work. Sleep does not liberate, but helps us recover so that we can work again, making it a part of the fixation in our modern lives on pure functioning.

Insomnia is altogether a different matter. Precisely because it is *not* recuperative, and because it unfolds within us a different sense of time and a world of thoughts beyond functionality, it has a highly liberating effect. Although Cioran stresses that it is necessary to pursue insomnia to its end, passing through the first stage, which is essentially one of unrest, anger, sadness, and other inflaming emotions. We have to cross over into the second stage of the second half of the night, in which our overtaxed mind becomes increasingly dulled. 'During bad nights,' writes Cioran, 'there comes a moment when you stop struggling, when you lay down your arms: a peace follows, an invisible triumph, the supreme reward after the pangs which have preceded it. *To accept* is the secret of limits. Nothing equals a fighter who renounces, nothing rivals the ecstasy of capitulation.'

This shows that Cioran had not only anticipated the therapeutic effect of sleep prevention for those suffering from depression. His 'ecstasy of capitulation' follows in the footsteps of the 'Dionysian' Nietzsche. But, as far as capitulation goes, Cioran forges his own path. And it is a completely unheroic one. After battling fruitlessly against his insomnia for seven years, Cioran decided he would no longer fight the dominant systems around him. His reasoning was, 'If you think long enough against someone or something, you become a prisoner of those thoughts and eventually come to love that slavery.' So, no more struggling, since that did not correspond to the emptiness Cioran was so familiar with through his insomnia. This has now been confirmed by brain research. It shows that sleep is not only regenerative, its memory-storing function also continues the activities of the day (inside our heads) and thus prevents us from entering the realm of emptiness that comes upon us after a long period of insomnia.

Instead, Cioran brings emptiness back down to the ground — the ground of non-existent facts. While Nietzsche still spoke of the giddiness we feel when we encounter emptiness, which requires us to have courage, Cioran stresses that emptiness is an 'abyss without vertigo' and courageous heroism is its complete opposite. Cioran warns of the danger of 'making the void a substitute for being', and exalting it, for example, to a special kind of bliss. He believes that in emptiness we are 'saved and unhappy for all time'. And for him, the Buddhists' nirvana is 'like suffocating, but in the gentlest way, at least'. Emptiness becomes bleak and trivial, an 'inner Sahara' — and thus thought through to the end, in a very real sense.

This is reflected in the fact that Cioran never makes any concrete suggestions as to how we can realise emptiness in our lives. Should we take some kind of drugs to keep us awake at night? No answer. Should we meditate? On this, Cioran says, 'To reflect means to realise that everything is impossible. To meditate means to elevate that realisation to the level of nobility.' That's not really much help, either. The Romanian thinker did not see himself as an adviser but as a philosopher of experience, scattering meaningless fragments and unable to help anyone. Not even himself: 'The longer it goes on, the more sceptical I become about my chances of dragging myself from one day to the next.' This, however, does fit in with the idea of the 'ecstasy of capitulation' described above. The exhilaration of helplessness is an essential aspect of emptiness.

Cioran experienced that exhilaration once more, in a very personal way, at the end of his life. He descended into mental derangement, in a similar way to Nietzsche, but for a couple of years less. Cioran, who had read countless books, refused to touch another one and constantly listened to music instead, although he had rarely done so earlier in his life. In his prime, he had advocated suicide as a possible way of escaping the meaninglessness of life. Later, suffering from dementia and dependent on the care of others, he no longer spoke of killing himself. Now his life really had become meaningless — and he no longer had any need for suicide.

3

Marching in Slow Step

the brainwaves of emptiness

From the works of these philosophers, we might already have guessed that there is more than one kind of emptiness. And we have examined the work of only a fraction of those who concerned themselves with this issue. We could name a whole host of others, not least of all Jean-Paul Sartre. After a life characterised by fear and uncertainty, he describes in his novel *Nausea* how emptiness can turn from enemy to friend — which also reflects the main thesis of this book:

'Nausea has not left me and I don't believe it will leave me so soon; but I no longer have to bear it, it is no longer an illness or a passing fit: it is I.'

Many meditation teachers, and other advocates of the so-called New Age, like to stress that there is only one kind of emptiness — namely 'their' kind. But, in fact, the brain can 'empty itself' in a great variety of ways. It can be still and quiet, as with yoga, meditation, or locked-in syndrome, or it can be loud and rhythmical with much motion, as with dancing, music, and sex. People can find their way to emptiness alone, in secluded solitude, or in a group, such as at a concert with a crowd of 10,000 fellow fans, or in the stands at a football match. An experience of emptiness can leave us refreshed and rested (which is presumably what most of us hope for), yet Emil Cioran's sleepless nights left him feeling drained and exhausted. Even highly trained Zen meditation practitioners say they are often left in pain, and in some cases with haemorrhoids, by their meditation sessions.

Of course, a Zen master would probably not want to be placed on the same level as a chanting football fan; and who would immediately accept that the emptiness of a foetus is comparable to the emptiness of an emergency patient who is close to death?

There are many different 'concepts' of emptiness, and this makes it difficult to define. Although it must be said that *something* is generally easier to define than *nothing*. Thus, when attempting to define emptiness, we cannot avoid focusing on the things that are *lacking*.

Neuronal fire protection: why the brain needs a fuse box

The human brain contains around 86 billion nerve cells, or neurons, whose structure looks a little like a clove tree. Clove trees form extremely fine branches, and the same is true of our neurons. Their projections can grow extremely long, stretching from one side of the brain to the other — and narrowing to a diameter of 0.1 micrometres. That's only a tiny fraction of the diameter of the fine twigs of a clove tree, but the filigree branching has very similar 'intentions' in each case.

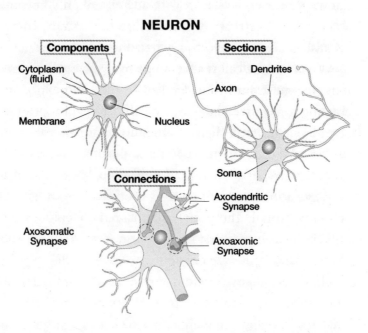

The clove tree has such fine branches to allow it to develop more leaves, creating more surface area for harvesting sunlight. Neurons do not gain any energy with their long projections, but they do gain better contact with other neurons. Just as a clove tree 'wants' to be struck by light waves, neurons 'want' to be touched by other neurons.

Those contacts generally take the form of synapses, which number in the hundreds of trillions — more even than the number of neurons. It's a figure with 14 zeros, and it shows the key role that connectivity plays in our brains. The synapses rely on messenger substances, also known as neurotransmitters, such as glutamate, dopamine, serotonin, and oxytocin. What is less well known, however, is that those transmitters only become active when electricity flows through the system. This is because neurons use electricity to transmit information.

When a nerve cell is at rest, positively charged sodium ions migrate from inside the cell to the cell membrane, creating a negative charge inside the cell and a positive charge outside it. When a signal arriving from another cell is strong enough to cross a particular threshold of excitation, the charges are reversed. The result is depolarisation, and the resting potential becomes an action potential. The positively charged ions flow from outside the cell to inside it, creating a temporary positive charge there. The neuron begins to fire according to the all-or-nothing-principle, meaning it is not activated just a little but completely. It sends its excitation charge travelling at high speed down special cell projections called axons to the synapses at their ends. Once there, the charge briefly triggers the release of neurotransmitters, before the electrical charge continues on its way. The excitation reaches the next neuron via special projections called dendrites. It then shoots down the next neuron's axon to the next synapse, where it continues on to a third neuron, and so on. In this way, a neural pathway is created, and it is along such pathways that neurons communicate with each other.

This neuron-synapse system sounds very simple, but in fact it is an extremely complex structure. Each individual neuron may be connected via hundreds or even many thousands of synapses to other neurons. Furthermore, when the brain transmits information, there is always some loss and 'conflicts'.

For example, a neuron's excitation can peter out and disappear into the void because it fails to reach the threshold value required to excite the next neuron. We now know that most synapses are pretty much 'weaklings'. This means each one alone does not have enough power to leave a sufficient impression on the next. They have to combine forces with other synapses and bundle their signals into the dendrites. This is similar to the way in which many tributaries combine their waters to create a mighty river flowing powerfully through the landscape.

What's more, the effect that the neurotransmitter substances have on the synapses is not necessarily always excitatory, and can often be inhibitory. So it is possible, and even common, for the excitation of one neuron to lead to the inhibition of another elsewhere. The American neurobiologist Neil Carson illustrated this with a vivid example.

Let's imagine I'm taking a hot casserole dish out of the oven to place it on the table for dinner. Suddenly, I realise my oven gloves aren't thick enough, and my fingers begin to feel the heat of the dish. My immediate reflex would be to drop the casserole to avoid burning my hands. The corresponding motor neurons in my spinal column even begin to fire, but they are immediately inhibited by an impulse from my brain saying, 'Don't drop it!' After all, it would make a real mess on the kitchen floor if I did.

Not to mention the fact that I would have nothing to eat for dinner.

In the heat of the moment, such rational thoughts probably don't even make it into my conscious mind, but they exist nonetheless. Our brains have created the relevant neural pathways with inhibitory interneurons in their synapses. Interneurons emit neurotransmitters that curb the excitation of the motor neurons, making them fire off less strongly. The result is that I grit my teeth, rush to the table, and put the casserole dish down, avoiding a culinary disaster.[1]

Carlson points out that his example is deliberately simple. We have the pain, the immediate reaction to that pain, and the suppression of that reaction. In real life, our actions are often far more structurally complex. Moreover, such inhibition can affect not only actions, but also thoughts. However, the principle is always the same: on the one hand, we have neural activity, and on the other, neural inhibition; the central nervous system is made up of 'fighters', which go off with all guns blazing, and 'inhibitors', which suppress firing. And this protects not only my kitchen floor from being splattered with hot beef bourguignon, but also my neurons from exhaustion and eventual death.

Meanwhile, the nerve cells of the cerebral cortex are designed in such a way that they will always fire, on principle. There are no inhibitors here; nature has designed this area of the brain to be a tireless thought pump, working both night and day. If it were left to its own devices, the cortex would continue to build up more and more electrical charge everywhere, until it all became too much. This would result in a massive discharge in the form of a seizure, overwhelming the brain's owner. The constant firing of such an unbridled cortex would provoke one epileptic fit after another, and this would mean the death of many neurons through complete burnout. The hyperactive cortex would eventually bring about its own demise.

To stop this happening, a fuse box has been installed — as in any regulation house — mainly in the form of the thalamus, with its inhibitory neurons and neurotransmitters. Making up the lion's share of the diencephalon, the interbrain, it is located directly beneath the cortex. From there, it features prominently in the decision about which signals should be let through to the cortex to cause excitation. This is why the thalamus is often called 'the gateway to the conscious mind', and as such it filters out what is 'meaningful' enough to be appreciated by the cerebral cortex. For our purposes, however, we are more interested in what it *does not* let through. If emptiness is to set in in the cortex, this can only happen via the thalamus and its strict filter. For this reason, the thalamus could also be called 'the gateway to emptiness'.

Incidentally, there is one animal in the natural world known for its particularly effective thalamus: the cat. And isn't that the very pet that has impressed us for millennia with its ability to doze and remain aware of its surroundings at the same time? Cats can sit around for hours on end, eyes closed, almost motionless, but not sleeping. At such times, their thalamus makes sure their cortex is not roused by external stimuli, while at the same time remaining ready to spring into action whenever an important stimulus, such as a mouse scurrying through the undergrowth, is detected. Such a cat is in a state of alert emptiness. And the fact that cats can often be heard purring in this state indicates that they experience it as being pleasant.

The wave theory of emptiness

The brain of a dozing cat is recognisable by its production of what we call alpha waves. The human brain is also capable of producing such waves, for example when we lie relaxing in the grass with our eyes closed. We may not be quite as skilled at this as a cat, but we can do it. However, what does it mean, when we say a brain produces alpha waves? And what are brainwaves anyway?

When several neurons in the brain fire at the same time, their individual activities combine into one brain activity. This leads to a particular wave pattern, which can be measured and displayed with an EEG (electroencephalogram). The pattern's oscillations may be extremely varied, ranging from delta waves with a frequency of less than 4 Hz to gamma waves with a frequency above 30 Hz; peaks can also vary greatly in size. The waves are like an orchestra conductor for the human brain. They control our level of alertness, our perceptions, and our memory formation.

A SIMPLE OSCILLATOR CIRCUIT

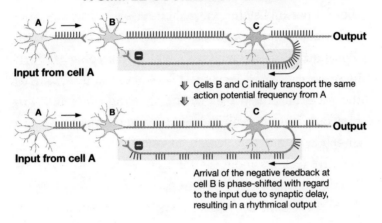

Cells B and C initially transport the same action potential frequency from A

Arrival of the negative feedback at cell B is phase-shifted with regard to the input due to synaptic delay, resulting in a rhythmical output

Production of waves in the brain

If the many billions of nerves cells and their connections in our brains were ordered in a chaotic way, they would also produce chaotic discharge patterns. Our subjective experiences would be completely disorganised and devoid of any content or concrete ideas. The fact that information can arise within the nerve tissue, allowing conscious content to be created, is thanks to the brain's ability to produce oscillations. Each wave is the sum of many individual discharge events within a defined network of neurons.

The illustration shows, for example, how the continuous firing of cell A is propagated, via cell B, to cell C. However, cell C also has inhibitory connections (minus signs in the illustration), which loop back to cell B. This negative feedback arrives — via a detour that involves crossing a synaptic cleft — with a slight time delay, such that the flow of excitation from B to C is broken. As shown in the next illustration, this creates pauses in the flow of excitation. An electrode placed over these cells would register oscillations in which each group of excitations would produce an electro-negative wave. Cell C now discharges rhythmically as an oscillator. Depending on the number and locations of the cell structures involved, various rhythms in various parts of the brain are created.

In real terms, this means that we can make assumptions about a person's state of consciousness by observing their brainwave patterns on an EEG screen. We can tell whether the subject is asleep and, if so, which sleep phase they are in, or we can see if the subject is awake, and if so, whether they are currently concentrating hard on something. However, it is not really possible to use brainwave patterns to tell whether the person in question has positive or negative feelings about their current activity or lack of it.

Many experiments have shown that certain classes of neuron can have a greater influence than others on the oscillation state of the network. An oscillation state exists when the body remains at rest: in that case, the cells of the motor cortex and connected areas oscillate with a frequency of 8–14 Hz, which is known as the sensorimotor rhythm (SMR). Interestingly, inhibitory neurons, which account for around 20 per cent of the nerves cells in the cortex, play a key role in this.

It should also be noted that the individual wave types can be superimposed on each other. Only alpha waves, which are typical of a relaxed, dozy but waking state, are unlikely to appear in conjunction with other wave patterns. This fact undoubtedly contributes to the feeling of pleasant emptiness we experience in that state. The alpha activity makes sure that the cell structures involved remain in an inhibited but receptive state. The low-frequency brainwaves of coma patients are also not usually overlaid by other, faster oscillations. But, perhaps surprisingly, it is often during sleep that our brainwaves really go wild.

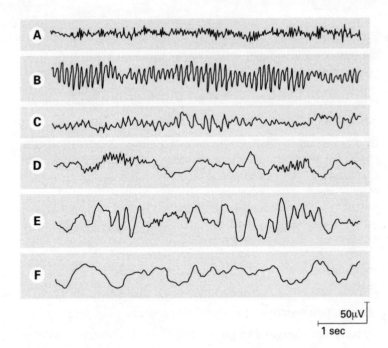

50μV

1 sec

Different wave types

A: *High-frequency beta and gamma waves (13–30 Hz and 30–100 Hz respectively).* Beta waves can have very diverse causes. They are observed during REM sleep, but also while awake and while under the influence of psychoactive drugs. Low-amplitude beta activity is indicative of active concentration or tension. Gamma waves, by contrast, appear not only during highly concentrated thinking and perception, and during intensive learning, but also during deep meditation of the kind described in Zen Buddhism as 'practising emptiness'. So we should not jump to the conclusion that the slower the brainwaves, the greater the emptiness.

B: *Alpha waves (8–12 Hz).* These waves are very harmonious and regular in structure. They are typical of a pleasantly relaxed waking state. They indicate an equilibrium, a balance between inhibition and excitation. This means that these waves often (but not always!) create good conditions for emptiness while awake. An example of a typical alpha-wave situation would be lying in a warm bath.

C: *Theta waves (3.5–7 Hz).* Mostly (but not always!) high amplitude. Occurrence increases with sleepiness and — with low amplitude — in the REM phase of sleep. But theta waves can also occur in the waking state. Depending on their location and amplitude, low-amplitude theta rhythms in the anterior sections of the brain in particular can signal a state of alertness. Recent research indicates that hippocampal theta waves may be the electrical expression of spatial perception.

D: *Sleep spindles (8–15 Hz) mixed with initial delta waves (0.5–3.5 Hz) in the left and right sections.* These low-frequency waves are characterised by a relatively large amplitude. They are a typical sign of deep sleep. Sleep researchers recently developed music whose frequency in the infrasound range stimulates the brain to produce delta waves. This induces an irresistible urge to sleep. In a study involving 170 patients, use of this delta-wave music was about as successful in treating insomnia as pharmaceutical sleeping drugs — with no side effects whatsoever.

E: *Deep sleep.* Delta waves mixed with high-amplitude theta waves (right, approx. frequency of 4–5 Hz).

F: *Coma.* Delta waves up to 1 Hz.

Can we sleep our way to emptiness?

Regardless of whether the man who later described himself as an 'actor behind a mask' really did suffer with ill health or was simply feigning it, the young Descartes managed to persuade his Jesuit school to let him stay in bed until noon because of his sickly constitution. And he continued the habit as an adult pursuing his philosophical studies. He loved to lie in bed and observe himself as he nodded off, dreamed, awoke, and dozed, half-sleeping, indulging in his own thoughts.

One such thought was: how can I be sure that my waking state is reality, but my dreams are not? A rather obvious question for someone who spent the greater part of his life in bed. His consideration of this question led

Descartes to conclude that everything can be doubted except the fact of doubting itself. This made him not a nihilist, but one of the most important rationalists. A life without thoughts, thus an emptiness of mind, was unimaginable for Descartes. And, more importantly, unbearable. Which is why he recommended treating children like patients with intellectual defects that must be driven out of them. The philosopher who loved to sleep became not a lover of emptiness but the complete opposite.

This was certainly not true of Emil Cioran. For seven interminably long years, sleep simply refused to come to him. Bed was not a place of relaxation, recuperation, or pleasant dreams, but one of struggle, tension, and swirling scraps of thoughts — and eventually the devastating but euphoric feeling of having lost the battle. This experience did not lead him down the same path as Descartes, who had felt the need to seek out ultimate certainties in the sea of similar dreams and waking thoughts. Cioran the insomniac rather felt no calling to anything at all, and, for him, there was only one certainty: the ceaseless thought machine in his head was not a blessing but a curse. Only when it could be switched off did existence become bearable. Perhaps not pleasant or worthwhile, but at least bearable — and human beings cannot expect any more than that.

At first sight, it may seem remarkable that it is not the sleep-loving philosopher but the insomniac who became the champion of emptiness. Even the Ancient Greeks described sleep as the half-brother of death, and so it should provide an ideal route to emptiness. Yet both the sleep-loving Descartes and brain science teach us that sleep allows us to enter a world that is so convincingly realistic that we can actually consider it real. This does not particularly smack of emptiness. And brain science confirms that there is a lot going on when we sleep.

Although we subjectively experience the slow waves in an EEG — characteristic of deep sleep, comas, and general anaesthesia — as emptiness, this experience is mostly a retrospective interpretation of the fact that we lose all sense of time in such states. It seems to us as if time passes 'in an instant' and so we later feel that nothing happened during that 'lost' time. In fact, the opposite is true.

During the first two to three hours of a night's sleep, which we are particularly apt to describe as 'empty', information is transferred from the hippocampus to the brain's long-term memory, in the cerebrum. Psychologists call this process 'memory consolidation'. It can be seen in an EEG, where it appears as rapid waves (known as 'ripples') originating from the hippocampus. These ripples are mostly superimposed on the slow waves of unconsciousness (mostly below 3 Hz), which are produced in the cerebrum. This doesn't mean that the two wave patterns work against each other; rather, the cerebral slow waves of unconsciousness are necessary to protect us from distracting thoughts and stimuli, thus freeing us up for the activities of the hippocampus. Furthermore, the mess of thoughts we have accumulated throughout the day is broken down in the form of adenosine phosphate. So our brains keep pretty busy while we're asleep. The fact that we subjectively feel recuperated after sleeping is connected with the physical processes of regeneration during the first phase of sleep, including the recovery of our immune system.

These superimposed brainwaves are absent in coma patients and in deep narcosis, so those states do not involve any hippocampus-controlled thought consolidation or formation of associative meaning. In such cases, we could speak of 'real' emptiness; 'real' because it's complete. Medics have identified 'islands' of information processing in the brains of coma patients and those in a persistent vegetative state, but these are usually very small and limited areas of the brain. We usually have no recollection of such processes because the hippocampus does not connect them to familiar contexts to create a meaningful whole.

More violent or pathological forms of unconsciousness, such as those caused by an electric shock, a blow to the head, or febrile convulsions, also create a state of complete emptiness. Negative memories, such as memories of loss or the violence suffered, as well as the rapid waves from the hippocampus associated with them, are deleted, at least for some time, and all that remain are the low-frequency deep-sleep waves of the cerebrum. Of course, there also remain the risks that are associated with this way of 'creating emptiness'. For this reason, we recommend other ways of immersing ourselves in emptiness.

In lockstep and happy: what sex, soccer, and military parades have in common

In fact, the kind of emptiness we experience in deep sleep, under general anaesthetic, in a coma, or in a persistent vegetative state has much in common with activities that are far less tranquil, indeed, even highly charged or hyperactive. These include epileptic seizures, ecstatic dancing in clubs, or orgasms. But they also include chanting in unison as part of a crowd or marching in step — as seen not only in historical reports about the Third Reich, but also in modern-day military-propaganda parades. The marchers at such events look positively enraptured, as if transported to another world, and certainly not unhappy or desperate, as we often assume from our modern individualist point of view. The fact is that when soldiers lose themselves in the lock-stepped marching of their corps, they become somewhat removed from this world, and thus achieve a little bit of emptiness.

This is because, in all of these activities, our brainwaves 'get into the rhythm'. Wide areas of the brain become synchronised, with the nerve cells dancing and singing in step over great distances. This means oscillation patterns are superimposed on each other, reinforcing them — i.e. increasing their amplitude. And the slower that shared rhythm is, the more alertness and consciousness fade into the background. The frequency of coma patients' brainwaves lies between about 1 to 2 Hz; during an epileptic seizure, deep sleep, or orgasm, they range from 1 to 4 Hz. Chanting in unison or dancing in time with a beat creates a synchronised, high-amplitude wave pattern rhythmically connected to the physical movement of those activities. This is usually similar to a theta rhythm, which is always significantly slower than that of beta waves as observed during highly focused thinking.

Most readers will be unsurprised to find that orgasms and dancing are on a level with ecstasy and emptiness. Yet they may be shocked to learn of the parallels with frenzied lock-step marching. But closer inspection reveals the compelling logic.

We know from personal experience, but also from experiments in the lab, that we will do almost anything to achieve an orgasm. Levels of the hormone responsible for sexual drive, dopamine, increase by 90 per cent in the brains of male rats whenever a sexually receptive female is within reach. We can safely assume similar levels in humans, although, so far, hormones released by the hypothalamus can only be measured indirectly (in the blood). If rats are given the opportunity to bring themselves to climax through self-stimulation, they take such frequent advantage of that opportunity that all other activities, such as eating, are totally suppressed.

This means rats are prepared to starve for the sake of those orgasms. All that just for a slow brain synchronisation of 1 to 4 Hz. Which humans can apparently experience while marching in step. No wonder participants love a good parade. Similarly, epileptic children will often wave their hands in front of their eyes to create a strobe effect, with its rhythmical interchange of darkness and light, to bring on a seizure.

All the phenomena described above involve the temporary synchronising of our brainwaves to achieve a state of cognitive emptiness. This is not the same thing as the motivational 'drive' orientation of the brain, which is always aimed at achieving some effect or other. Emptiness is not something that represents a positive stimulus. You can't want emptiness in the same way that you can want a piece of chocolate or the respect of your fellow human beings. It would lose its very status as emptiness if it were able to be a classic object of our will.

However, our brain sees things differently. It is able to see emptiness as so desirable that it will make us do anything to get it. So there must be something to emptiness. To investigate what that might be, it makes sense to begin by examining the architecture of the brain.

4

Beyond the Defence Mechanism

the brain areas of emptiness

As we have seen, the wave patterns of the brain arise out of the interaction between active and inhibitory neurons. Those neurons are not uniformly distributed throughout our grey matter, but are organised into anatomically or functionally distinct units or areas. Some parts of the brain, such as the hippocampus or the thalamus, are very old in evolutionary terms. Others, such as our cerebral cortex, are evolutionarily young. Accordingly, the various areas of the brain differ greatly in their tasks and functions. This doesn't mean that the 'upper' (cognitive) and 'lower' (emotional and motivational) areas of the brain run counter to each other. Instead, they complement and even depend on each other.

THE MOST IMPORTANT AREAS OF THE BRAIN FOR AN UNDERSTANDING OF EMPTINESS

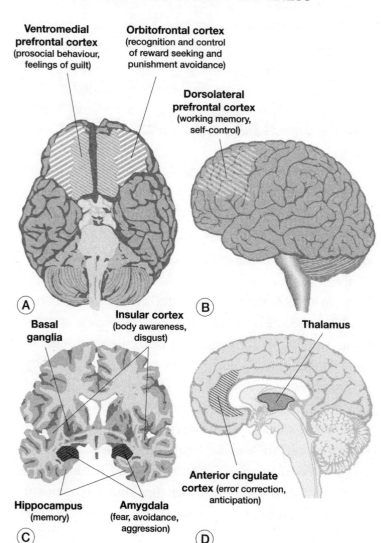

Some animals, such as elephants, rely on physical bulk to ward off danger; as the world's largest modern land animals, they have little to fear from predators. Chameleons and peppered moths, by contrast, rely on camouflage, and the clownfish nestles among the poisonous tentacles of the magnificent sea anemone. As the German Enlightenment philosopher Johann Gottfried von Herder put it, humans are 'defective creatures': too physically weak to survive in nature, we have weak teeth, eyes, and ears, lack fur, and — in comparison with our close relatives such as gorillas or Neanderthals — have pretty puny musculature. Yet we humans do have a brain that, although it makes up only 2 per cent of our total body mass, accounts for 20 per cent of our body's overall metabolic activity.

With the aid of this calorie-guzzling mini-colossus, humans are able to compensate for their physical defects. It allows us to learn to recognise dangers early and defuse them with more intelligent countermeasures. Thanks to their brains, early humans were able to defy famine, drought, enemies, and competitors for resources, which played an important part in the survival of our species. Today, in the industrialised world at least, those skills are no longer vital for survival, but since evolution requires a certain amount of time to adjust to altered circumstances, our brains still work almost as they have always done, namely as sensors of danger whose job it is to help guide us safely through the vicissitudes of life.

In this context, the social psychologist Martin Seligman speaks of the 'catastrophic brain'. What this means in real terms is that our brains have evolved always to expect the worst. We tend to focus first on what could go wrong, and not what does go well. This is why our brain's defence system is particularly pronounced — and particularly sophisticated in its internal workings.

The amygdalae play a key role in this system. These are two almond-shaped clusters of nuclei (the word comes ultimately from the Greek *amygdale*, 'almond nut') nestled deep in the brain, one in the left temporal lobe, the other in the right, just in front of the hippocampus. The central part they play in our defence system becomes clear when they are damaged or deactivated in the lab. Birds captured in the wild, which would normally attempt to fly away in panic, suddenly become calm and placid when their amygdalae are no longer functional. And laboratory rats will even seek out the company of sedated cats and sniff at them inquisitively.

In the journal *Current Biology*, scientists at the University of Iowa published a report of a woman whose amygdalae were destroyed by disease.[1] She is still able to experience happiness and sadness, but she can no longer feel fear. Before falling ill, she had a pronounced phobia of snakes and spiders, but can now even touch such animals. 'She was simply overcome with curiosity,' explains Justin Feinstein, the lead scientist in the study. When researchers visited a notoriously 'haunted' house with the woman, or watched a horror movie with her, she showed less reaction than the scientists who were supposed to be studying her responses. When she was asked to fill in a computerised emotion diary, she was happy to comply, but her log provided about as much excitement to read as the minutes of a philatelists' congress. After all, what remains of a personal journal if all fears and worries are absent from it? This begs the question: how could this woman even survive, without the feelings of fear that prompt us to avoid dangerous situations?

In Tübingen, we had a young patient who was brought to us by his mother because of his 'contradictory' behaviour. In traffic, whenever a car approached from the right, he barely took notice of it and simply walked out into the road. If a car approached from the left, he would stop at the curb and wait for it to pass. The boy's teachers complained that he would sometimes seem completely uninterested, but in sports lessons would sometimes act like a reckless daredevil, only to return the next minute to his usual fearful behaviour. An MRI scan of his brain revealed that while his left amygdala was intact, the right one had been destroyed by a febrile infection (unlike the two hemispheres of the cerebrum, the nerve pathways from the sensory organs to the amygdalae do not cross each other). This meant that he no longer had any fear of things approaching from the right — and so the right half of his body reacted like that of a fearless psychopath.

People whose amygdalae are unusually large display a pronounced and easily triggered defence reaction. Scientists have observed this link in autistic infants, which is probably explained by the fact that such children see their fellow human beings as generally alarming and so are unable to draw any sense of security from interpersonal contacts. A study carried out by University College London points in a similar direction.[2] This research didn't examine people with autism, but rather focused on adults who had spent their childhood in unsafe family environments. Their amygdalae were also typically enlarged. And the same is true of voters who regularly choose candidates who espouse conservative social values. When people feel they are unsafe or under threat, they tend to want to cling on to the status quo and to that which is familiar and therefore comforting for them. In this aspect, conservative voters and autistic children are remarkably similar.

However, we should take care not to paint an incorrect picture of the amygdalae on the basis of this connection between the organ and political orientation. Indeed, the connection is found in animals as well as humans. The amygdala is an extremely ancient organ and functions like an alarm system. Within milliseconds, it evaluates the level of danger in a given situation and, if it is deemed to be high enough, a command is sent via the hypothalamus and the pituitary gland to the adrenal glands to begin pumping out stress hormones. The result is the classic fight-or-flight response: blood pressure and pulse rate rise, muscles tense, digestion and pain perception are rolled back. If it weren't for this pattern of responses our species would have already died out.

But how do the amygdalae know when to sound the alarm? This can happen in two ways: one quick, rough, and prone to error; the other slow, but resulting from very precise analysis. Both paths begin with the thalamus, the central switchboard for messages arriving from our sensory organs. It can pass signals on in two ways: directly and indirectly.

The first alternative is aptly described by the American neuroscientist Joseph LeDoux as 'quick and dirty'.[3] Here, the thalamus passes a quick sketch of the sensory stimuli directly to the outer nucleus of the amygdala, where a decision is made, on the basis of innate mechanisms and learned knowledge, about whether the inner nucleus should be informed or not. If so, the actual defence response is launched from there. All this happens without us having to think about it. Or even being *able* to think about it. A typical 'quick and dirty' reaction is when we catch sight of an indistinct shadow in the corner of our eye and immediately flinch, and our pulse rate skyrockets.

The second pathway, dubbed the 'high road' of cognitive processing by LeDoux, is more complicated. In this alternative, the stimulus first passes from the thalamus to the cortex and the hippocampus, where the sensory perceptions that triggered it are analysed. A more nuanced evaluation of the information takes place in the sensory regions of the neocortex, for example to distinguish the lighter footsteps of a woman approaching from the sound of the heavier gait of a man, or to differentiate between a harmless bumblebee and a potentially dangerous wasp. Or recognising that the menacing shadow in the corner is our own, and not that of an intruder. Additionally, the hippocampus also introduces conscious memories of unpleasant or frightening past experiences into the mix. Which is why a person who has once been stung by a bee and had an allergic reaction to it will react differently to the sight and sound of a yellow-and-black insect buzzing round them than an apiarist.

The hippocampus can calm our defence system, and one way in which it can do so is by recognising that the levels of stress hormones coursing through our bodies are too high and then sending signals to the thalamus to scale back their production. However, the hippocampus can also rev the system up again. For instance, by providing us with the information that the face opposite, which we initially categorised as unfamiliar, is actually that of a cousin. We might calm down when we recognise that we know the individual in question. On the other hand, this information may have precisely the opposite effect if we cannot stand the cousin in question. In that case, the defence system and the amygdala will be kickstarted into action.

Another region of the brain that's important in the defence system is the insula. It uses involuntary changes, such as an increase in heart rate and muscle tension, to inform us of our current emotional state. The insula also plays a part in pain perception, and not just when we are in the dentist's chair or involved in an accident. A Canadian-American research team investigated the brains of people with a pronounced fear of mathematics.[4] They found that when such people were given a maths problem to solve, their posterior insula fired off as if they had just broken a leg.

The anterior insula, the front part of the region, by contrast, is a control centre for empathy, which need not necessarily always be focused on others, but can also be self-focused, as in the case of self-pity. Without this area of the brain, we would be unable to assess whether a voice — including our own — sounds sad or not, or to judge whether a face staring at us is contorted with anger or pain.

Most of the connections from all those motivational areas end up in various parts of the prefrontal cortex, which is huge in human beings, and which also forms part of the defence system. It is located in the front part of the brain and is made up of a large number of different areas and structures serving various executive functions. They coordinate the information arriving from the anterior part of the brain with recorded data from the past to create a flow of behaviour that leads us unerringly to rewards — with respect to our organism's aims and expectations — or helps us avoid punishment.

The short-term memory and self-control are among the executive functions of the prefrontal cortex, and are therefore also prerequisites for our ability to create emptiness within ourselves. We need the self-control provided by the prefrontal cortex at least for the conscious triggering and introduction of emptiness of thoughts, and it is not until the experience of emptiness kicks in that the frontal cortex is 'cut off' from the other control areas of the brain. Without the prefrontal cortex, emptiness would overrun us in an uncontrolled way.

The cingulate gyrus, on the other hand, is involved in the storage of negative memories. It is the negative stooge to the hippocampus. When we do something that turns out not to live up to our expectations, the cingulate gyrus breaks the chain of action in order to initiate a new one. This mechanism does not work so well in hyperactive people. That is why they often persist with behaviour even when it does more harm than good — and drives the people around them up the wall.

All the areas described above require a system to keep them alert and awake while they are 'on the job'. This is provided by the ascending reticular activating system (ARAS). It is difficult to define structurally, stretching in the form of a 'column of cells' through the central part of the midbrain to the thalamus, with which it cooperates closely. We have already met the thalamus as the 'gateway to the conscious mind', letting through only powerful or important information. The firing rate of the ARAS determines how open that gateway is. Powerful stimuli cause an immediate increase in frequency — and we are suddenly wide awake. Working together with the thalamus, the ARAS controls our level of alertness, although this process should not be thought of as 'pure' activation, like the switching on of a desk lamp.

The thalamic reticular nucleus is mainly home to inhibitory neurons, whose slow wave patterns reach up to higher levels to leave our waking cortex oscillating in a calm and relaxed alpha-wave rhythm. If, for example, an external acoustic stimulus that is deemed to be of importance arrives, the thalamus and the ARAS will let it through, 'up' to the auditory centre in the cortex. In order for us to perceive that stimulus consciously, the thalamus keeps the other areas of the cortex oscillating in an alpha-wave rhythm to prevent them from 'jamming' the signal and creating interference noise, which would hinder perception of the acoustic signal or even prevent it altogether. We could say the thalamus prevents the brain from getting bogged down in multitasking.

It's not difficult to imagine that this mechanism plays

a major part in the creation of emptiness. When the thalamus 'closes the gate', and allows next to nothing to pass on up to the cortex, that part of the brain will remain quiet, but this is now a very rare occurrence. In prehistoric times, humans often experienced periods when they could 'switch off': waiting behind a bush for their prey to turn up, walking 30 kilometres to the next watering hole, or simply staring into the campfire at night. It is reasonable to assume that their brains were full of alpha waves at such times. But in the industrial age, such periods have become almost non-existent.

Whether we are at work or at leisure, there is always something of significance going on. Your smartphone pings as a new message arrives, the TV screen flickers across the room, you rush off to the store to buy that thing that's only on special offer for one more day, or you gather round the water cooler with your co-workers to moan about the boss. Any of those is significant enough for the thalamus to open its sluice gates and for the ARAS to send excitation signals up to the cortex, where high-frequency wave patterns will then gain the upper hand.

What's more, everyday stimuli are often associated with anxieties, which means the amygdalae also come into play. The times may be over when we needed to fear being attacked by sabre-toothed tigers, struck by lightning, poisoned by berries, or killed by infected wounds. But those fears have simply been replaced by others: fear of losing our smartphone, missing out on that special offer, being fired by the boss, or failing a maths test. All this means that our defence system is constantly active, despite the fact that it was developed by evolution as a survival strategy in extreme situations.

It's not surprising that we are often afraid of emptiness, since it can represent a break with and a painful end to the lives we have become accustomed to. That is why our catastrophic brains activate our defence systems when confronted with the idea. It is no coincidence that many children resist going to sleep, because bedtime means being ripped out of the flow sensory perceptions that is the waking world.

A desire for emptiness is nonetheless connected to the fact that it gives our defence system a chance to rest. Nothing and no one likes to live in a permanent state of emergency!

The state of emptiness should then result in increased activity in the brain's reward centre, since the two are in an antagonistic-inhibitory relationship with one another. The reward system is made up of a series of regions and nerve connections, such as the nucleus accumbens in the basal forebrain and the basal ganglia deep inside the brain. This network is what motivates us, controlling our urge to be active. A reward strengthens the system's connections between the areas for movement ('I'll do this and that …') and the areas for the hoped-for consequences ('… to get this and that'), and the neurotransmitter dopamine forms the glue for those connections.

However, we still do not know whether this system is particularly active when we experience emptiness. What we do know, is that we can still experience something as positive (a reward) even when no more dopamine is present. Lack of dopamine simply means we can no longer initiate activity to pursue that positive stimulus. We are still able to *be* happy, but can no longer do anything to *become* happy. This would then be a kind of happiness that is not associated with activity, but rather a contemplative, purely observational happiness. This explains why the brains of completely paralysed locked-in and Parkinson's patients still have a functioning reward system, as we will see later.

Furthermore, many people report having profited in some way from experiencing a state of emptiness. For example, some may say they feel like they have 'refilled their fuel tanks', others cite gaining creative impulses and new perspectives. Scientific research into meditation points in a similar direction. Of course, those results should be viewed with caution because we don't know whether meditation makes people creative or creative people are particularly drawn to meditation. Yet the simple fact that people associate emptiness with positive effects can trigger increased activity in our reward centres, causing us to view emptiness as something desirable and making us want to create a positive kind of emptiness.

The question remains of how best to attain this state. As the 'natural enemy' of emptiness, the defence system must be scaled back, since it leads to alertness and meaning. As we have seen, this can happen in several key places in the brain. And that in turn means we can approach emptiness in several different ways.

5

Default-mode Network

the brain on autopilot

The weather was warm in the summer of 1928, when the Scottish microbiologist Alexander Fleming left a Petri dish of bacteria culture out on his laboratory table. When he discovered this later, mould had already started to spread through the Petri dish. Fleming was about to throw it away when he noticed that the bacteria in the colony had not reproduced in the vicinity of the mould. This was the birth of penicillin — the name given to the substance in the mould that was making life so hard for the bacteria.

Some 70 years later, the Pfizer company first tested a substance called sildenafil on male patients and were initially as disappointed as Fleming had been when he discovered his mould. This was because the substance did not turn out to have the desired effect in treating the heart condition angina pectoris. But when the test patients were asked to return the tablets of sildenafil they still had, they refused. These men had finally found a solution to their impotence problems. Pfizer decided to take advantage of this unexpected side effect, and marketed the drug as a treatment for erectile dysfunction. That was the birth of Viagra.

The importance of chance in the history of science should not be underestimated. As in the case of Marcus Raichle and his colleagues at Washington University in St Louis in the mid-1990s, when they put a few test subjects in a PET scanner, as was the fashion among brain scientists at that time. A positron-emission tomography scan can measure where activity is going on inside our skulls, much more accurately than an EEG. This technique produces multiple images of the brain, like an onion when it is sliced rather than diced. The clever part is that patients are injected with a radioactive solution before their scan. This accumulates in areas of the brain where there is a lot of metabolic activity, which is in turn an indication of neuronal activity. And this information is then displayed as an image on a monitor.

Raichle and his team were carrying out a rather unspectacular experiment, aimed more at testing the machine than their subjects' brains. Volunteers were placed inside the big donut-shaped scanner and asked to follow a dot with their eyes as it moved around on a screen. In the time between those exercises, they were expected to simply lie there and relax until the dot started bouncing around on the monitor again. It was a simple alertness and concentration test, nothing more. The biggest problem should have been simply staying awake between those bouts of eyeball acrobatics.

But that did not turn out to be a problem. Quite the opposite, in fact. When Raichle examined the images of his volunteers' brains, he found the signs he expected of neurons busily firing off in the frontal cortex and the frontal eye fields of the brain as they followed the dot. What he was surprised to find was that the subjects' brains did not shift down a gear when the screen was blank. Metabolic activity in the frontal eye fields took a dive, but in other areas it increased considerably. These people's brains were working about as hard as before, but in a different way, in different areas. Despite the fact that they were receiving very few stimuli from the environment — such as the steady hum of the machine itself — to process. So what was going on?

An analysis of the images revealed that four areas of the brain began firing synchronously and showing increased activity during the rest periods:

1. the posterior part of the cingulate gyrus, which is involved in the associative connection of memory and emotion
2. the posterior part of the parietal lobes, where visual perceptions are combined to create an overall impression
3. the hippocampus, as the central location for the consolidation of visual memories
4. the medial prefrontal cortex, which is responsible for decision-making and evaluating probability.

This rather accidental result piqued Raichle's curiosity. He combed through past studies and found that his subjects were not abnormal in showing brain activity during resting: other researchers had noticed the similar phenomena. 'They didn't pay any attention to it,' explains Raichle, 'because the unoccupied brain was not on the scientific agenda at that time.' This discovery of this synchronised activity has now entered the scientific canon and has been dubbed the default-mode network (DMN). Raichle describes it as 'an organised, baseline default mode of brain function that is suspended during specific goal-directed behaviours'.[1] This implies the converse is also true: that our brains do not just shut down when we let our minds wander. Goal-directed behaviours are curbed while in other areas activity is ramped up. This explains why the brain's metabolic rate is only 5 per cent lower in the default mode than during concentrated activity or observation, or when we are solving a maths problem.

The distribution of brain activity in the default-mode network suggests a base state of readiness in the brain; and the regions of the brain that remain active, as described above, suggest that this state of readiness takes the form of a kind of mental rehearsal of certain functions. For this purpose, the brain busily draws on its stock of memories and sensory impressions, and associates them in different ways; decisions are anticipated and sounded out by the brain to predict their future influence on and significance for a possible solution to the problem being considered. In Raichle's study, the subjects' brains were presumably assessing the task of following the dot on the screen. But the task could have been anything: acoustic recognition of syllables, for example. The principle is the same: prompted by tasks it predicts will come up in the immediate future, the brain goes into standby mode, allowing it to spring back into action without a long preparation period as soon as the task crops up. This saves both energy and time.

But what happens when the task at hand changes? What would Raichle's test subjects have done if they had suddenly been presented with not an optical task, but an acoustic one after the default phase? The simple answer is: they would have had trouble. Various studies have shown that such a situation results in the precise opposite of time and energy saving, and it takes a large number of neuronal impulses for the brain to adjust to the new situation. Such a change can eventually even result in the release of stress hormones. This isn't a disaster, but it does show that our brains do not like 'cold starts', preferring to settle into a certain operating mode, which allows them to solve tasks with the highest possible level of efficiency.

In our everyday lives, this allows us to concentrate on a task without losing the thread if we stop for a while to do nothing. That is much more difficult if we break our concentration to complete another task that requires concentration, such as checking our phone messages. So, in the work environment, it makes much more sense to interrupt tasks with real breaks: by taking a walk for half an hour, or staring vacantly out of the window for a few minutes.

Raichle's default mode is still somewhat removed from emptiness. A better comparison would be a computer working in offline mode, with no wi-fi access, using the disconnected time to defragment its hard drive. It cannot open any fresh websites, but it reorders its data files, making it better equipped to complete future tasks. Raichle's experiments show that the path to emptiness is not an automatic reaction, and our thought pumps do not just switch themselves off when we cut the power and reduce stimulation. However, we must bear in mind that Raichle's subjects were pausing between bouts of concentrated activity. They were expecting to have to start performing the task again very soon. But what happens to a brain that is not in that state of expectation, and is left alone with no task to complete and no prospect of one?

More attention for the important things in life

This was the question that interested Kalina Christoff, a neuropsychologist at the University of British Columbia.[2] Her test subjects were simply told to make themselves comfortable and do nothing, but to remain awake. In other words, they were asked to let their minds wander, to just daydream. With no aim, no stated purpose, no pressure to solve any problems or do anything at all. Their brains were scanned during this process using magnetic-resonance imaging, which measures the changes in blood flow to the various parts of the brain.

This experiment showed once again that activity increased in the areas of the brain identified by Raichle, as evidenced by the increased blood flow to those regions. But that was not all. Christoff and her team discovered that the so-called executive network of the brain also become active during mind-wandering. These areas are primarily located in the prefrontal cortex. This network is usually mentioned by researchers in association with such processes as goal-setting and prioritisation, conscious attention control, or the targeted initiation and coordination of activity. That sounds like a far cry from emptiness.

Interestingly, Christoff also discovered that the executive network becomes particularly active when we are unaware of the fact that our minds are wandering. This is a strong indication that the areas of the prefrontal cortex are less concerned with specific individual goals than with higher questions. 'When our minds wander,' says the Vancouver-based neuropsychologist, 'we may not achieve our immediate goal, for example, following a lecture or the plot of a book. But it is possible that we are using that time to deal with more important questions in our lives.'

This assessment has since been confirmed by studies carried out by other scientists. In one such study, Christoff's fellow neuropsychologists Randy Buckner and Daniel Carroll of Harvard University compared the brain activity in default mode with that found in studies of the brain in other waking states.[3] 'Large parts of the default network are identical to the parts of the brain that have been identified as being particularly active during all types of self-projection.'

Self-projection goes on when we call up memories of past events and re-evaluate them, or when we envision future events and evaluate the probability of their actually occurring. It also includes our ability to adopt the viewpoint of others, or view ourselves from the outside — we might imagine ourselves in the place of a partner when we tell them that we are leaving them for someone else. People with autism lack this ability, so it is no surprise that scientists have been unable to identify a consistent default pattern in their brains.

Leonhard Schilbach, a neurologist at the Max Planck Institute of Psychiatry in Munich, also found indications that 'self-referential thoughts' are turned over in a wandering mind.[4] His results showed that the brains of test subjects who were asked to think about themselves displayed similar patterns of activity to those of subjects who were asked to do nothing and think about nothing in particular. The Australian neuroscientist Ben Harrison, on the other hand, found patterns typical of the default mode in the brains of subjects who were asked to imagine themselves in the place of someone in trouble, and think about ways to escape that situation.[5] Altruism, empathy, and the default-mode network all appear to be on the same 'wavelength'.

In any case, it seems that what the mind is doing when it wanders is principally changing our perspective to imagine what others think about us and what we would do and how we would feel if we were someone else. Our own ego is then no longer an acting subject, but an object of observation. This could certainly be described as a kind of 'partial emptiness', since it involves abandoning reactive thinking oriented towards a specific problem or sensual impression.

However, this theory cannot be tested on the level of brain physiology. We could ask people what is going on in their mind while it is wandering, but they would have to return to a reactive state of consciousness in order to answer, thus breaking off direct contact with the default-mode network. They would then at best be able to provide only memories of their daydreams that had already been processed in terms of coherence and logicality — and these would not necessarily be the same as the actual processes in default mode.

The negative effects of mind-wandering in everyday life

We cannot necessarily assume that drifting into autopilot mode will make us happy. That might be the case when we are staring absentmindedly out of the window of a train, or listening to the hum of the MRI machine during a scan, far removed from the pressures of everyday life. But when we are in the midst of day-to-day life, daydreaming can be a very negative experience.

A team of researchers at Harvard University recruited almost 2,300 male and female volunteer subjects for an experiment carried out while they were going about their daily lives.[6] They were repeatedly contacted via their smartphones, where they had to mark on a screen what they were doing at that moment, and how happy or unhappy that activity was making them feel. In this way, the researchers were able to gather a total of 250,000 snapshots from the lives of their test subjects, and a spontaneous evaluation of those moments.

DAYDREAMING AND THOUGHT COMPLEXITY

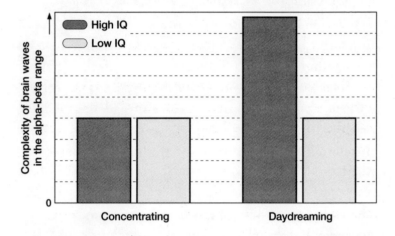

Daydreaming, intelligence, and emptiness

This illustration shows the complexity (unpredictability) of the brain activity of highly intelligent and less intelligent individuals, who were given two different tasks to complete: one was simply to let their minds wander, i.e. to indulge in daydreaming; the other was to concentrate exclusively on recognising certain letters displayed within a rapid series of other letters.[7]

We see that the complexity of the electrical processes in the brains of highly intelligent subjects at rest and while daydreaming is far greater than that of less intelligent subjects. The level of complexity was calculated using mathematical algorithms from the field of nonlinear dynamics (better known as 'chaos theory'). This involves determining how often a particular wave pattern is repeated during the completion of a task, and consequently how predictable it is. A high level of complexity means low predictability of brainwave patterns.

During mind-wandering, thoughts slip away from the task at hand or from an external stimulus, and move aimlessly back and forth among personal memories, plans for the future, and pure ideas. Our study shows that this process is much more variable and unpredictable in highly intelligent individuals.

In general, however, there is a correlation between mind-wandering and deficient cognitive performance, especially during completion of complex and longer-lasting tasks, as the subject's focus moves away from those tasks towards more easily accessible memories. This happens more quickly and more often in less intelligent individuals, and they tend to remain for longer in the world of their daydreams.

Mind-wandering and emptiness (both at the level of the brain and also on a mental-cognitive level) are barely compatible, if at all. True emptiness requires a shift away from personal memories and problems.

The results showed that the volunteers spent 47 per cent of their waking lives daydreaming. Irrespective of whether they were washing the dishes, hanging out the washing, travelling to work, or surfing the internet — their thoughts tended to wander. For almost a third of the time, their minds were busy with something other than what they were supposed to be doing. This appears to show that we spend a large proportion of our lives in a sort of non-presence.

That sounds more like depressing alienation than blissful rapture. And indeed, such mind-wanderings do have a negative effect on our quality of life. The study described above showed that those who were particularly apt to let their minds wander were less happy than those who did so less frequently. Drifting off into daydreams was felt to be just as dissatisfying as doing a monotonous, repetitive job or surfing aimlessly on the internet.

Dissatisfaction with daydreams is based in part on the fact that it keeps us from working and stops us from achieving our aims. If you indulge in daydreams instead of studying for an exam, the result will be frustration due to a failure to carry out a planned activity — and a reduced chance of passing that exam. Then, daydreaming is no longer experienced as a restful break from the pressures of everyday life, but as a cause of agitation or — as Zen

Buddhists put it — the 'grasshopper mind', which jumps from thought to thought and hinders our efforts to complete the challenges of our everyday lives.

However, mind-wandering can also make us unhappy in a more direct way, since it is often self-referential. And those who are not at peace with themselves, but rather agonise about what other people think of them, about whether they are accepted or ignored by those around them, whether they are loved or despised, will be less socially confident — and lack of confidence is not exactly known to help in achieving happiness in life.

Whatever the case, there are clearly disadvantages to repeatedly zoning out of everyday life and letting our minds wander. The temporary emptiness this creates in individual parts of the brain is also only marginal, because our thought pumps don't shut down altogether, and may even be more susceptible to influences from outside than otherwise. To get closer to emptiness, it is necessary to do more than just zone out of day-to-day life and indulge in daydreams.

6

Senselessly Happy

*what happens to us when
nothing happens*

My eyes open and I see — nothing. Absolute darkness. My automatic reflex is to prick up my ears, but once again — nothing. Absolute silence, except for the sound of my own breathing. I move my hands to feel where I am, and this is a very different experience — there is something surrounding me! Above me, below me, beside me. It feels like wood. My sense of smell joins in and signals the odour of soil — and now, finally, a terrible realisation begins to take over: I have been buried alive!

I begin to panic, to sweat; my breathing becomes shallow and my pulse begins to race — all precisely the wrong reactions in an airless hole in the ground. I fight to regain my composure, but my feeling of horror just won't subside. I scream for help, while at the same time, thoughts flash through my brain. Why me? Yes, I was ill. And I've fainted often in the past. But any doctor should have been able to tell that I'm still alive!

And then, at last, comes release: I feel myself growing sleepy, and I can barely make a sound anymore. Everything grows calm. The lack of oxygen is taking its toll. I'm suffocating to death.

People have been burying their dead in the ground for at least 50,000 years. We cannot know whether they, too, feared being buried alive. But since the Middle Ages at least, when the influence of Christianity led to a gradual rise in the popularity of burial over cremation, this fear, known to doctors as taphephobia, has been spreading. And like all phobias, it is basically unfounded: it is very rare for excavations of graveyards to reveal skeletons in twisted positions or to find scratches on the insides of coffins — signs of a struggle for survival beneath the earth. There are no reliable figures showing how common it was for people wrongly thought to be dead to be lowered into the grave, but we can be certain that it makes more sense to be afraid of getting a fatal electric shock from your kitchen toaster than to fear being buried alive.

Despite this, taphephobia became a widespread theme in the world literature of the 19th century. Edgar Allan Poe and Gottfried Keller, for example, both wrote about it in their stories and poems. The American author was in no doubt when he wrote: 'To be buried while alive is, beyond question, the most terrific of those extremes which has ever fallen to the lot of mere mortality.' The Danish writer Hans Christian Andersen left a note every night on his bedside table reading, 'I only appear to be dead.' And even the ever so enlightened and hardnosed philosopher

Arthur Schopenhauer left a provision in his will that he should not be buried until six days after his death.

Right into the 1950s, some German mortuaries still had bells installed that people could ring for attention if they awoke after having been erroneously declared dead. Anglo-American inventors marketed 'safety coffins', which gave live-burial victims 72 hours after the funeral to raise a signal above ground. There is not a single documented case of that ever having happened. Nevertheless, taphephobia has not become any less common. Telling ghost stories around a campfire at night, tales of being buried alive are still a sure-fire way of making listeners' faces turn pale with fear, despite all the zombie apocalypses, vampire attacks, and chainsaw massacres modern audiences have already witnessed.

But what if you knew that you had not been doomed to die in your coffin, and could rely on being rescued? What if you were not only cooped up in a small space, robbed of your senses of sight and hearing, but also of your sense of touch and your sense of your own body? Unimaginable, you say? Perhaps you should think again. Because in such a sarcophagus of sensory deprivation, you can experience an emptiness in which your perception of the external world is extinguished, and the problem of fear along with it.

Floating: from hallucination to deep relaxation
In the 1950s, North American neuropsychologists began intensively investigating the question of how the brain reacts when it is cut off from the barrage of information is usually receives via our sensory organs. Until then,

the dominant theory was that such sensory deprivation would have an effect comparable to pulling the plug on an electric lawnmower: the brain would switch to sleep mode. But there was no real proof to back up this theory. That gap in research was destined to be filled first by the Canadian psychologist Donald Hebb.[1]

Together with his student Walter Bexton, Hebb recruited participants for an unusual experiment at McGill University in Montreal. Subjects were offered 20 dollars — a relatively large amount of money at the time — for every day they spent doing absolutely nothing. They were asked to stay in a soundproof room, which was insulated against noise from the outside, but with the constant hum of an air-conditioning unit audible inside. This meant the subjects were not able to hear the sounds made by their own bodies breathing and swallowing, or the sound of their own hearts beating. They wore goggles with heavily frosted glass, isolating them visually, while very soft furnishings, cotton gloves on their hands, and cardboard tubes on their arms meant their sense of touch was very limited. Apart from this, the experiment did not appear to make great demands on the subjects: do nothing and earn good money while doing it — Hebb had no problem finding enough recruits for his study.

However, those recruits would soon regret their decision. Most aborted the experiment after just two days, and none managed to endure this 'senseless' environment for longer than a week. The subjects found the experience simply unbearable. Soon after the start of the trial, they began to have difficulty concentrating. Their minds would constantly wander, causing them to score abysmally on the cognitive tests they were required to complete periodically. They found themselves bombarded by memories that raged unfiltered and unbridled into their conscious minds. Many subjects reported hearing voices that were not really there, and some even heard music. Others saw wallpaper patterns or entire scenes, where they encountered dogs or children, or felt they were being shot with pellets. The strangest hallucinations were experienced by one subject, who saw a procession of squirrels in snowshoes with sacks over their shoulders marching across the room. That even trumps the visions of another volunteer, who saw prehistoric monsters walking through a jungle.

After the test subjects abandoned the experiment, they had great trouble reintegrating into the world. Some felt for several hours afterwards that the outline and size of the objects and people around them were constantly changing. It appeared the theory of the switched-off brain was history. Now it seemed human beings could even lose their minds when deprived of sensory stimuli — not exactly a desirable state, of course. Hebb concluded that the neurons in the sensory regions of the brain will begin to occupy themselves when they are starved of stimulation, resulting in strange fantasies. Cognitive function then deteriorates. So, the Canadian researcher concluded further, such deprivation should be avoided, and thus education should aim to provide children with as much sensory input as possible.

The US neurophysiologist John Lilly was not prepared to take that conclusion at face value. Even in his time as a student, the medic and biophysicist was known for his unusual experiments. Once, during an anatomical human dissection, he stretched an entire intestinal tract across the length of a room to determine its actual length with certainty. The other students and teachers watched in amazement as the young Lilly laid out the human gut — which can famously be more than six times longer than its owner — but nobody tried to stop him, recognising the fact that he was not doing it just for a jape, but to gain a scientifically correct measurement. Lilly published his first academic paper while still studying for his doctorate. It was an investigation into the effects on the human body of two particular amino acids. The experiment involved Lilly going on a diet that included next to no protein and taking only measured amounts of the amino acids under investigation. He became increasingly weak and eventually delirious as a result. Nonetheless, he later claimed that he had not found the experience unpleasant.

In 1954, while working for the National Institute of Mental Health in Maryland, Lilly developed the first isolation tank. It allowed subjects to be without not only their senses of hearing and sight, but also touch, and their sense of their own body.[2] Its construction was reminiscent of a bathtub in a darkened, soundproof cabin. Its internal dimensions were around two metres in length and 50 centimetres in width, to avoid the risk of claustrophobia. Test subjects would lie in extremely salty water, so that the resultant buoyancy allowed them to float without any part of their body touching the tank. The water temperature was maintained at just under 35°C, which is about the same as the exterior temperature of human skin. This ensured that the subjects felt neither particularly cold nor warm. The temperature of the water and the lack of external stimuli aimed to allow subjects to lose the sense of the physical boundaries of their own bodies — a state described by Lilly as 'floating'. The water was not deep, between 25 and 30 centimetres, allowing subjects to sit up and end their floating experience at any time.

Lilly first tested the tank on himself. Initially, it made his skin sting, but he solved the problem by reducing the concentration of salt to a level that would still allow him to float but no longer irritate his skin. On a mental level, however, he experienced no such problems. He did not feel sleepy, nor did he lose his mind. Even after several hours of floating. Rather, he reported feeling an 'enormous but silent joy', 'completely new inner experiences', and 'altered states of consciousness'. He further sums up the experience, 'Nowhere else can such a deep level of muscle relaxation be achieved as while floating.' For this reason, Lilly suggested floating could be an effective alternative in pain therapy. And he was later proven correct, when floating in salt solution became an accepted part of treatment for rheumatic pain.

Yet the academic world remained sceptical of his claims of altered consciousness. Lilly was known as a maverick who liked to take his experimentation to extremes, especially when investigating mental states, and so it was assumed that he slipped into the salty water with a biased view. His co-workers and students, who later acted as floating test subjects, would also have had a fundamental interest in exploring new psychological experiences. But how would 'average people', who did not have such interests, react to being detached from their senses? Would they even be prepared to get into that dark, poky isolation tank?

Emptiness can only become pleasant when fear is removed

Like Lilly, Peter Suedfeld is also a researcher who likes to take the less comfortable route. Born in Hungary, he fled the horrors of the Nazi occupation with forged, 'Christianised' papers, and so even at the young age of barely ten, he became familiar with the capacity of human beings to cope with extreme situations and traumatic events. From the 1970s onwards, while working at the University of British Columbia in Vancouver, this topic became the focus of his psychological research. He has investigated the psychological strain experienced and coping mechanisms used by researchers in the Antarctic, prisoners in solitary confinement, mountaineers, military generals, and astronauts.

Suedfeld was interested in finding out how survivors who had experienced the Holocaust in their youth coped with their traumatic experiences.[3] He interviewed such people and analysed existing academic literature on the subject. Suedfeld found that, despite their ordeal, many Holocaust survivors managed to lead creative, successful, and fulfilled later lives. They were socially integrated, had families, and maintained stable, trust-based friendships. 'There was no evidence of a paralysing trauma,' the Canadian psychologist writes.

He theorised that these people's personality traits had interacted with their experiences in the Holocaust in such a way that they were able to negotiate their new environment well. For one thing, they wanted to prove with success in their professional and private lives that they were not the 'parasites' the Nazi regime had labelled them as being. In some cases, Suedfeld suggested that the survivors' near-death experiences had resulted in a desire to leave a long-term legacy for future generations. Of course, such a view had its detractors in the academic world; and, indeed, it should be borne in mind that Suedfeld investigated only 'successful Canadian copers'. His study did not take into consideration those victims whose experiences in the Holocaust led them into depression or alcoholism, or had already led to their death — and their numbers were not small.

But Suedfeld's methods were not limited to interviews and analyses of other people's results. He carried out his own empirical research. He was also fascinated by the work of Donald Hebb, whose test subjects spent days in a kind of isolation cell, cut off from the normal sensory world. Yet Suedfeld felt unable to accept Hebb's results. His doubts centred on the question of why human beings, who are capable of coping with all kinds of extreme situations, should be unable to bear a situation in which there are no sensory experiences and therefore no real dangers. This led him to re-examine the setup of experiments such as Hebb's closely.

This revealed that test subjects had been made to feel quite fearful about the various experiments. For example, they were required to sign a release forms protecting scientists from any claims for compensation should something go wrong in an experiment. They were also given and informed about 'panic buttons' that they could press at any time to abandon certain experiments. Obviously, such measures put the idea into the test subjects' minds that something could go wrong. Also, they were not given the chance to examine the test environment before they entered it. All this must already have activated the subjects' defence systems to relatively high levels: they were on the alert and therefore not in the right frame of mind to just accept the sensory deprivation as it came.

Not to mention the fact that the sensory deprivation experienced by test subjects was far from total. For instance, Hebb did not remove acoustic stimuli by creating silence, but with the constant hum of an air-conditioning unit. Nor did he counter the effect of gravity. Hebb's subjects received enough external stimuli for them to be constantly aware that they were locked up in a kind of isolation cell. It goes without saying that such knowledge is not exactly conducive to calming a person's fears.

This led Suedfeld to repeat Hebb's experiment but under altered conditions.[4] His setup was similar to Lilly's, with a large, darkened, soundproof tank in which subjects could float weightlessly. The water was kept at skin temperature and only smelled vaguely of salt, if at all, so that the subjects' senses of touch and smell had little stimulation. They also had no reason to fear tipping over to one side in the water, as the salt solution buoyed their bodies in a stable, supine position. The same phenomenon allows tourists to float easily in the Dead Sea without having to move a muscle.

Suedfeld minimised his subjects' natural fear of the unknown by allowing them to examine the tank beforehand and providing them with extensive information about how it worked. Also, there was no panic button; instead they were told they could leave the tank at any time, and the floating experiment would be over in 60 to 90 minutes in any case. Additionally, Suedfeld gave his sensory-deprivation experiment the non-threatening title 'Restricted Environmental Stimulation Therapy', which he abbreviated to the cosy-sounding acronym REST, with all its associations of relaxation, recuperation, and regeneration. Thus, the subjects went into the experiment with the feeling that they were about to have a pleasant and relaxing experience, rather than with the fear of being locked in an isolation cell.

Suedfeld's concept was successful. Only 5 per cent of his test subjects aborted the experiment early, and most of those did so because they were unable to stand the initial itchy feeling on their skin after entering the salty water. However, there were no reports of psychological problems such as hallucinations or anxiety attacks. Many of the subjects even had to be coaxed out of the tank at the end of their floating session, because they did not want the deep state of relaxation they found themselves in to end. A small number fell asleep, while most reported feeling awake and enjoying following their 'flow of consciousness'. This sounds very similar to Lilly's accounts, but it can be assumed Suedfeld's subjects were not familiar with those reports.

Other scientists have examined the wave patterns produced by the brains of people floating in sensory-deprivation tanks. These measurements showed that beta waves are significantly reduced after about 40 minutes, while theta waves increase strongly. As a reminder: these are the low-frequency waves produced by the brain as it falls asleep, indicating a state researchers like to call the brain's 'twilight state'. Only, most subjects did not fall asleep in the tank. This indicates that during sensory deprivation, we conserve a state that is usually experienced as pleasant but fleeting. Imagine lying in a hammock on a warm spring day, relaxed and dozing, no longer quite awake but not quite unconscious, either. In such a state, our brain activity is almost certainly dominated by theta waves. Under normal circumstances, we can only enjoy that state for a moment — but it can be extended to many minutes' duration in a floating tank.

In this simple but effective way, Suedfeld managed to make his subjects' experience of sensory deprivation a pleasant one. First, he removed their fear of the isolation tank, thereby deactivating the defence circuits in their brains. Second, he created a state of sensory deprivation that was nigh-on total. This enabled him to prove that when our brains are detached from the body's senses, they are not forced desperately to produce a replacement world, but can adapt to the situation perfectly adequately and 'go with the flow'.

In the 1980s and 1990s, Suedfeld carried out much more comprehensive testing, which showed that repeated sessions in the tank have a positive effect on memory, creative problem-solving, and other cognitive abilities.[5]

For example, subjects became more imaginative and productive in brainstorming sessions when their brains were regularly detached from their physical senses. Suedfeld's main explanation for this effect was that a brain that is liberated from the control of the senses is better at coming up with novel solutions. As if the brain were saying, 'If nobody is going to show me the way to go, I'd better come up with something myself.'

Yet when Suedfeld and his students tested whether the procedure would have a positive effect on the improvisation abilities of jazz musicians, they were in for a surprise. They divided a set of musicians into two groups, one of which simply attended regular practice sessions for a period of four weeks, while the others also underwent sessions of sense-less floating in isolation tanks. Suedfeld had expected the latter group to be the better improvisers. But when Suedfeld played recordings of their sessions to a top jazz expert, he was unable to hear any difference between the two groups' improvising abilities.

Spontaneous creativity had not improved in the 'floaters'. That group were found to have improved their physical playing skills, however. Their playing was simply better. They made fewer mistakes, the saxophonists' embouchure was cleaner, the guitarists' fingering was more precise. This may have been because the sensory-deprivation sessions improved their ability to filter out all the extraneous stimuli and concentrate better while playing, but Suedfeld was unable to provide a definitive explanation for this phenomenon. Emptiness often produces inexplicable phenomena.

The sixth sense: our feeling for our body

Those floating tank experiments offered convincing proof of the fact that it is easier for human beings to enter a subjectively pleasant state of emptiness if their senses are largely deactivated, and that perception of the dimensions, position, and movement of the body is key to this. Scientists call this sense 'proprioception', and it bears closer examination, as it differs from the body's other senses in many ways.

The first detailed consideration of the way we perceive the world around us is found in the treatise *De Anima* (*On the Soul*) by Aristotle (384–322 BC). His identification of five senses — sight, hearing, smell, taste, and touch — was to dominate the worlds of medicine, biology, and philosophy for centuries. It did not occur to the Ancient Greek philosopher to ask himself what gave him the ability to remain standing or lift a cup of wine without dropping it — and none of the scholars who came after him considered the point, either. That alone is enough to show that human beings are apt to concentrate on their perception of the external world as provided by those five senses. The fact that we have an inner world that we can also sense, is often overlooked.

It was not until the 19th century that doctors became aware of the organs of equilibrium we have in our inner ear, although they had long been aware of the phenomenon of dizziness. Our sense of balance is not only responsible for keeping us from falling over, it also controls our eye and head movements. Without it, humans would be unable to keep their eyes focused while turning around. Our sense of the movement and position of our bodies was also only recognised in the 1830s, by the Scottish anatomist Charles Bell. Initially, just as a general notion: 'We use the limbs without being conscious.'

Later, his idea took on more a concrete form: 'I called this consciousness of muscular exertion a sixth sense ... When a blind man, or a man with his eyes shut, stands upright ... by what means is it that he maintains the erect position? ... It is obvious that he has a sense by which he knows the inclination of his body ... In truth, we stand by so fine an exercise of this power, and the muscles are, from habit, directed with so much precision and with an effort so slight, that we do not know how we stand. But if we attempt to walk on a narrow ledge ... we become then subject to apprehension: the actions of the muscles are, as it were, magnified and demonstrative of the degree in which they are excited.'[6]

It was Bell's fellow physiologist, the Englishman Charles Sherrington, who first coined the word 'proprioception' to describe this muscle sense. It comes from the Latin words *proprius* (own) and *recipere* (receive, take). However, the two British scientists did not manage to persuade the world of the existence of this sense. In fact, it did not really gain traction until it was popularised by the English neurologist Henry Head, who indicated that it was an existential issue — in the true sense of the word — which he described as 'a model of ourselves'.[7]

Our visual and acoustic sense organs are clearly distinguishable, but the situation is more complex in the case of our proprioceptive sense, as it is the product of several different receptors. We have sensory cells that 'measure' the position of our joints, for example, and something called the Ruffini corpuscles, which are mechanoreceptors found not only in our joints but also our skin. They are sensitive to pressure and stretch, and can be activated by swelling or tensed muscles beneath the skin, even when we are doing nothing. Our breathing, our heartbeat, and the bloated feeling in our stomach after a blow-out meal are all registered by our Ruffini corpuscles, which are named after the Italian anatomist Angelo Ruffini who first described them. This means they play an important part in how (well) we feel in our own skin, so to speak.

Another key part in our proprioception is played by our muscle and tendon spindles, the latter of which are also called the Golgi organs. Muscle spindles are found within the body of our muscles and are aligned parallel to the muscle fibres. And just like muscle fibres, the ends of the muscle spindles can contract — while their midsection (the 'equatorial region') is non-contractile. Around that section is a sensory nerve fibre, which is distorted in shape when the muscle contracts, and which send a message to the central nervous system when that happens. The muscle spindles are *stretch* receptors, and that's how they differ from the Golgi organs.

Those receptors are also spindle-shaped and are found in the tendons — that is, where the muscle attaches to the bone. Unlike the muscle spindles, they are not aligned parallel to the muscle fibres, but in rows behind them. This enables them to sense changes in muscle tension and send signals to the central nervous system accordingly: they are *tension* receptors. Their sensitivity threshold is higher than that of the muscle spindles; they are not activated until there is a considerable build-up of tension in the muscle.

In principle, muscle tension can be built up in one of two ways: due to active tensing or due to passive stretching of the muscle. When we kick a football with our lower leg, the Golgi organs are activated in both the front *and* back or our upper leg. Those at the front spring into action because the muscles there are tensing, while those at the back are activated because the kicking motion means the muscles there become stretched. This means the central nervous system can send signals to the hamstring muscles at the back of the thighs to relax so that they are not injured as we score that goal. This normally occurs without the involvement of our conscious mind; such defensive reflexes prefer to take a shortcut via the spinal cord.

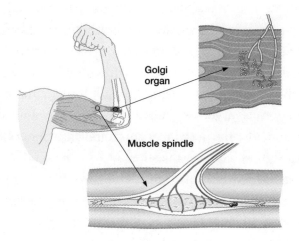

Muscle spindles and Golgi organs are among the central pillars of our proprioceptive sense. They inform the central nervous system about tension or stretch in our muscles.

The self and the will disappear with the sixth sense

It would be a mistake, however, to view the muscle and tendon spindles as nothing more than a mechanism to protect our muscles. They provide overall information about the amount of tension and stretch in our muscles. The importance of this in enabling us to control the movements of our body is demonstrated by the rare cases of people who have lost their sense of proprioception through illness. One such case is that of a butcher on the British Channel Island of Jersey, named Ian Waterman.[8]

At the age of 19, Ian contracted a viral infection that threw his immune system so seriously out of kilter that his own antibodies began to attack the cells of his peripheral nervous system that are responsible for the senses of touch and proprioception. His ability to feel temperature and pain remained intact, and Ian was still able to move — but all that was of little help to him. His loss of muscle perception meant he could no longer control his movements, because his brain was not receiving any of the feedback it needs for the fine adjustment of its orders.

Above his neck, everything was still functioning more or less normally, and so Ian did not lose the ability to speak. But from the neck down, his muscles no longer sent out any signals. This soon resulted in Ian's brain no longer sending out impulses to move. His *will* to do anything was paralysed, and, eventually, this formerly strapping young man no longer knew how to send a signal to move a finger or bend a knee.

If Ian didn't monitor the position of his limbs visually, they would sometimes even move of their own accord. Once, while Ian was being washed by a nurse, his hands brushed over her breasts. This was completely unintentional on Ian's part, but the nurse refused to believe it was so, and slapped him across the face. At the time, Ian was horrified, although he was later able to laugh about his undeserved punishment. After all, he wouldn't have been able to gain any pleasure from his involuntary 'groping' of the nurse's breasts, even if he had wanted to, since he had lost his sense of touch.

Ian's doctors were stumped by his case. Which is not surprising, since the first academic paper explaining what has now become known as 'acute sensory neuropathy syndrome' was published in 1980 — nearly ten years after Ian contracted his viral illness. He was told he would never walk again, let alone work, and was sent home. But he was not willing to accept that fate. Faced with such a bleak prognosis, his will was reawakened. He decided to work on improving his situation and spent every waking second thinking about movement and reminding his brain that it was, in principle at least, still able to trigger movements. He placed himself on a strict rehab regime, learning to control his movements using his visual sense. He made endless paper-clip chains and repeatedly practised putting on his socks, though it initially took him 20 minutes — per sock.

He sometimes suffered serious setbacks: for example, the time he was given a plastic cup of water to quench his thirst. By that time, Ian had relearned how to move a glass to his mouth and drink from it. But when he tried that manoeuvre with the plastic cup, he simply squashed it with his hand, because he had misjudged the strength of his grip. Ian's reaction was to go to the supermarket, buy several catering packs of plastic cups, and practise with them at home until he had relearned that skill, too.

What happens if our inner sense is switched off?

The importance of proprioception is also shown by some stroke patients. Their condition is the result of damage caused by a burst or blocked blood vessel in the brain. The damage often affects the nerves that transport perceptual information from the arms or legs to the cerebral cortex. As a result, such patients are no longer able to feel anything in those limbs and tend not to use them, although they usually retain some or all of the ability to *move* those extremities. Like Ian Waterman initially, they are paralysed because they can no longer control the movement of their limbs, although their motor function remains more or less intact.

The illustration shows a patient with a paralysed right hand, undergoing 'constraint-induced movement therapy' developed by Edward Taub. This involves restraining the patient's functional left arm for several weeks, forcing the patient to use his compromised right hand for all movements. This enables the patient to overcome the learnt non-use of his hand and reactivate the areas of the left hemisphere of the brain that are responsible for controlling the right hand. The proprioception lacking on that side of the body is replaced by healthy senses such as the ability to see motion and successful movement.

Eventually, Ian's efforts paid off. He learned to walk again, although even today his gait appears somewhat ungainly, since he has to control his steps mainly using visual signals. He also learned to complete other tasks, such as sticking a stamp on a letter, but has to do so by licking the envelope rather than the stamp, which could stick to his tongue, where he would not be able to remove it with his eye-controlled hands. He even passed the civil-service selection procedure and worked for 12 years for the government. There's no doubt about it — the man from Jersey provides an impressive example of the brain's plasticity and its ability to cope even with the total loss of such an important sensory system. It must be said, however, that Ian is an exception: science knows of no other patient so successfully compensating for a total loss of proprioception.

At the time of writing, Ian was running his own company, has been married three times, and very few people he interacts with are even aware of his condition. Though this is partly because Ian has never really found the right words to describe the state he lives in. This is no coincidence. A person who loses their sense of hearing is deaf; a blind person is one who has lost their sense of sight. But what about someone who has lost their sense of proprioception? It's clear he has lost the ability to sense the amount of stretch and tension in his muscles and the orientation of his body in space, and this leaves him unable to control his movements in the normal way. But that is only one aspect of a total proprioceptive loss.

Another aspect has to do with the fact that this inner

sense is also active in healthy people when they are *not* moving. In principle, there is always tension in muscles — it is what we call muscle tone — and the level of that tension correlates with the situation we find ourselves in. When we are afraid or stressed, that tension will be much greater, since the body is prepared for a fight-or-flight response. Ian's loss of proprioception means he can no longer feel any of that. He can no longer sense when he is under stress, which naturally means his defence system is activated much more weakly.

For people with an intact sense of proprioception, it can be difficult to calm down after exposure to highly stressful situations. This is because their muscles continue to be tense, and that tension signals to the brain that the danger has not yet passed. Ian, by contrast, is relatively quick to calm down after a stressful situation. He reports that, although his condition means he finds himself in dangerous situations more often than others, he does not experience those situations as intensely as other people. Although his life has become far more difficult, his ability to get stressed about those difficulties has become much less, which, in turn, makes his condition much more manageable.

But there is yet another aspect to proprioception, involving our ability to recognise the boundary between ourselves and the outside world. It tells us, 'I am here, and over there is another person. Here — me; there — not-me.' As well as the muscle and tendon spindles, the Ruffini corpuscles described earlier may also have a part to play in this, as they register the tension in our skin, and therefore also the extent of our bodies in space. Whatever their role, the signals they send out provide us with the certainty that we are living inside our own body, and that our body is distinct from the rest of the world.

Of course, this distinction is not made by our sense of proprioception itself, but in the organ where the signals from the proprioceptive system are perceived and analysed. In other words, in the brain. Researchers at University College in London have discovered that the right posterior insula plays a central role in self-perception and the structure of our body image.[9]

When a rubber glove becomes a body part

In their study, the British researchers used a variation of an experiment known as the Rubber Hand Illusion, which is one of the most fascinating phenomena in perception research. Test subjects place their left hand on a table and their right hand on their lap, under the table. A rubber right hand is positioned on the table in place of their hidden one. When the rubber right hand and the hidden real hand are then stroked in exactly the same way, 80 per cent of test subjects will report the illusion of feeling as if the rubber hand is actually their own. At some point, they will even begin to feel as if their hand is being stroked when in fact only the rubber hand is 'receiving the stimulus' and their real hand his not being stroked at all. This shows that the subject's brain has integrated an artificial object into its image of the body.

There are many possible variations on this experiment, using a different concealment technique (such as a cardboard wall) or without concealment (with both the subject's hands on the table next to the rubber hand), and the illusion still works. The important point is that, under certain circumstances, we are willing to view a foreign, lifeless object as part of our own body. And this can even go so far that subjects will reflexively pull their hand away when they see someone is about to jab the rubber hand with a needle.

Using positron-emission tomography, the British scientists observed the activity in their test subjects' brains during the Rubber Hand Illusion. This showed that when the rubber hand is perceived as just a rubber hand, the somatosensory cortex shows increased activity. This makes sense, since that is the part of the cerebral cortex that deals with tactile signals received from the receptors in the skin. But in subjects who have already integrated the rubber glove into their body image as their actual hand, a different part of the brain springs into action — namely, the right posterior insula. Accordingly, this is where an important part of our self-perception is located. 'The right posterior insula underpins the subjective experience of body-ownership,' explains the lead scientist, Manos Tsakiris. Perhaps it would be more precise to speak of a *claim* of body-ownership when a rubber hand is integrated into our own mental body image.

However, Ian Waterman makes no such claims in this respect. Because his proprioceptors no longer work, his insula cannot lay claim to a rubber hand, or his own hand, or any other body part for that matter. He does not have a mental body image, has no feel for such an image, and also therefore has no certitude of himself as a unit with clearly delimited boundaries to the world he exists in. He still remembers how he felt back when he was bedridden in hospital: 'My body was gone, I didn't really exist anymore.' It was only when he looked down at his body that he became aware of his own continued existence. When he closed his eyes, he felt as if he had disappeared once more into oblivion.

This is the reason he still prefers to sleep with the light on, because waking up in the dark would leave him lost, in the truest sense of the word. He would not even be able to change his sleeping position, because he would not be able to coordinate the necessary movements. Once, a well-meaning doctor advised him to sleep with a pocket torch next to his bed, so that he could sleep at least the majority of the time in the dark. But how could that help him? How can someone without a sense of touch or a feeling for movement find a torch in the dark? Ian ignored the advice and carried on sleeping with the light on.

This doesn't mean that Ian experienced these times of feeling lost and disembodied as unhappy. Even just shortly after he became ill, he repeatedly experienced moments when he didn't want to emerge from that state. 'You just want to lie there.' This is because the body's defence system has an opportunity to rest when its muscles are not sending out stress signals; and when the distinct self dissipates, its problems dissipate along with it. In a nutshell: when perception of the body's inner life is lost, inner peace can take hold.

A fellow sufferer of Ian Waterman's, known in academic

literature as GL, compared her situation with someone living on a sailing ship who has two choices: to remain below deck and go wherever the ship goes; or to make the effort to go up on deck and — again with great effort — steer it. At first, you go up on deck regularly to take control, but since this involves great effort (controlling movement using only visual perception without the aid of proprioception is extremely hard work!), you start to wonder whether the effort is worth it. After all, the ship can sail itself, only, without going on deck, there is no way to influence where the voyage takes you. So GL opted to remain below deck more and more often, and eventually remained there permanently. This is why she, unlike Ian, remains unable to walk and has to use a wheelchair. She has now also lost the ability to speak because her condition has spread to the receptors above the neck.

In many ways, GL is now like the locked-in patients we will meet later in this book. But does that make her less happy than Ian, who reactivated his will and was able to return to our world? As wilful creatures, we tend to believe she must be. But a certain experiment taught me early on that we should not be quite so quick to jump to that conclusion.

Emptiness is also a question of attitude

I was a visiting researcher at Rockefeller University in New York when I decided to experiment on myself with curare. The famous arrow poison used by some indigenous people of South America was attracting a lot of attention in psychology and physiology at the time because of its part in disproving a cornerstone of those sciences. That cornerstone was the belief that there is a strict division between learned behaviour and cognition on the one hand, and learning to control the processes of the autonomous nervous system — such as heartrate and excretion of stomach juices — on the other.

Researchers injected laboratory rats with curare, to completely paralyse them and exclude any possible influence of 'learning by doing'. Their astonishing findings were that despite being paralysed, the rodents still learned to raise their own heartrate at will. All that was needed was to reward them with an electrical stimulus to the pleasure centres of their brains every time their pulse rate increased. And the same was true of bodily functions such as increasing blood pressure, increasing blood flow to the kidneys, and changing brainwave frequency. Some of the animals even learned to control the blood flow to individual ears, making one appear slightly bluish in colour, while the other remained white.

Since rats and humans are rather similar when it comes to the structure of their circulatory and nervous systems, it seemed reasonable to assume *Homo sapiens* would be able to do this, too. It was surmised that humans can control their own autonomous bodily functions without the use of motor activity and without having to develop a specific behaviour pattern. As a result, many researchers, including myself, believed it would only be a short time until unhealthy or pathological physical and mental processes, such as gastritis or depression, could be simply 'unlearned'.

We know now that this is not the case, but that hope was widespread at that time. So I injected myself with curare to test the hypothesis that paralysis results in a loss of the urge that drives wilful, instrumental learning. I took a dose that would have been fatal under normal circumstances, since it causes total paralysis, even of the respiratory organs. To stop me from expiring in that way, I had an anaesthetist friend at my side to ventilate me artificially.

After just a few seconds following the injections, my body totally collapsed. The arrow poison works by blocking the neurotransmitter acetylcholine, which activates muscle contraction. This results in paralysis that lasts for as long as the curare molecules continue to block those docking points. That is what I felt immediately: the tension flowed out of my body like air escaping from a balloon. But I did not feel fear.

On the contrary, I felt a wave of profound relaxation spreading through me that I could not escape. It simply swept over me. If I was thinking anything else at the time, I no longer remember what it was. But I know I was not worried, as I knew the ventilation machines were reliable. I was not brooding over any problems, and no fragmentary thoughts were flashing through my mind. All was strangely calm.

As I realised the effect of the curare was beginning to abate, and muscle tension began to return — and breathing along with it — my feeling was not one of relief.

I was certainly not the first scientist to experiment on himself with curare. In 1944, for example, the English doctor and chemist Frederick Prescott did so and described the total paralysis, inability to breathe, and fear of suffocating as 'a very unpleasant experience'. In 1946, an American anaesthetist called Scott Smith subjected himself to several injections of curare with different doses without developing any panic reaction at all. He was even able to report in detail afterwards that he had remained fully conscious and able to feel pain, despite being paralysed. Neither he nor Prescott described a feeling of profound relaxation.

This difference is easily explained by the different circumstances of our experiments. Prescott had to be ventilated using a kind of handheld bellows, which is not exactly conducive to allaying fears of suffocation. And Smith was too fixated on recording his condition to sink into a deep state of relaxation. By contrast, I had complete faith in my anaesthetist (I would not carry out the experiment today, as my faith in the medical profession has been shaken too often in the course of my long life!), and I was more attuned to individual experience than scientific discovery.

We saw in the context of experiments in floatation tanks how important the subject's attitude to opening up to emptiness can be. It is no different with curare experiments. Those who wish to keep a clear mind, or those who fear it, had better not flirt with emptiness. This will only lead to frustration at not achieving emptiness, or problems when they do.

7

Training for Emptiness

why is a mouse when it spins

We have ascertained that certain conditions must be met for us to achieve a pleasant experience of emptiness. The brain needs to create certain wave patterns, and the body's defence system must be at rest. The flood of signals from our sensory-perception system needs to be at a very low ebb, and, in particular, proprioceptive self-perception should be at peace. All this is impossible to achieve without a positive, unfearful attitude to emptiness, and trust in the process that leads us to it. But which of such processes can we trust?

We have already considered some of them. Philosophy can help establish the necessary basic outlook and offer a little understanding of what emptiness really means.

Floatation tanks can provide a way to unshackle ourselves from the flood of stimuli from our sensory systems. Modern tanks no longer resemble dark coffins, instead reminding us more of nuts, which represent the origins of life. In California, they are about as common as sunbeds are in darker climes, and some people even have their own tank installed at home in their living room or bedroom. The ideal length of a floating session is between an hour and an-hour-and-a-half — easily included in a normal weekly routine, since it doesn't have to be done every day.

Appropriate relaxation techniques such as autogenic training or progressive muscle relaxation as developed by Edmund Jacobson are certainly helpful in achieving emptiness. They are proven to reduce muscle tension and therefore reduce the deluge of stimuli from the proprioceptive senses. Their effect in this respect is not as violent as that of curare, but, on the other hand, they do not require the use of a ventilation machine. However, these methods can only achieve the 'emptiness effect' if practitioners become skilled enough in them to progress beyond the stage of just practising. Anyone who is occupied with chanting 'my belly is full of warm bread rolls' or 'my pulse rate is pleasantly relaxed, powerful, and regular' still has a long way to go before achieving emptiness.

Another approach would be to target not the body's proprioceptors, but the part of the brain their signals are sent to for analysis and interpretation — the insula. That is a practicable alternative, as we have been able to show in Tübingen. The insula is able to learn, and one of the things it can learn is to ignore proprioceptive signals, at least to a certain extent.

Away from the island: how to find inner peace

The insula is buried below the temporal lobes and other

parts of the brain, and so it was not discovered until the late 18th century. Naturally, its evolutionary history goes much farther back than that. This is reflected in the development of foetuses in the womb: while other 'lobes' of the brain, such as the cerebral cortex, continue to grow as a foetus develops, the insular cortex stops growing when it reaches about an inch across. But in terms of the wide range of functions it performs, the insula is a giant.

Signals from virtually all sensory channels are processed and interpreted in the insula. Taste, for example, which we know can prompt a violent reaction of disgust — that reaction originates mainly in the insula. By the same token, it can also interrupt an irresistible attraction. According to one American study, many smokers who suffer damage to the insula due to a stroke lose their addiction. As Nasir Naqvi of Columbia University in New York explains, 'they simply had no more urge to smoke'.[1]

The insula is also the place in the brain where pain is given its emotional dimension. Some people panic at the sight of an ingrown toenail — and when that happens, we can be pretty certain that their insula is firing off quite violently. At the other end of the scale, we have stories like that of the legendary Manchester City goalkeeper who finished a game even after he broke his neck in a collision with another player. We must assume his insula was firing rather weakly.

English researchers irritated the skin of their test subjects with the pepper-spray agent capsaicin to measure not only the intensity of the pain it caused, but also the arousal of the subjects' insulae. They found that the more intense the reported pain, the greater the activity of the insula; the less intensive the complaint of pain, the lower the activity in the insula.

The insula is particularly well connected to the messages that come from within the body. It plays a major role in feelings like hunger, thirst, shortage of breath, satiety, and nausea. It lets us know when the walls of our stomach or bladder are distended, making us feel full or need the toilet. As already mentioned, it is in the insula that the proprioceptive lines of communication converge. If your neck muscles contract, your heart is beating in your mouth, and your breathing switches to panting mode, the insula takes careful note of all this and sends out the message: 'Uh oh! You're scared! Do something!'

A brain region that has such basal functions needs to be dependable, which means it must not be easily influenced. Nonetheless, the insula is certainly not an island, cut off from all external influences.

At our institute in Tübingen, we were able to condition the front part of the insula (the anterior insula) in a targeted way, using neurofeedback. This involves monitoring blood flow to the brain with the aid of magnetic-resonance imaging or another visualisation technology while a computer converts those signals into a symbolic representation on a monitor screen that the patient can see and easily understand. This representation might be a colour-coded thermometer, for instance, which moves into a red zone at the top, or it may be a plane that flies over an obstacle when blood flow to the insula increases.

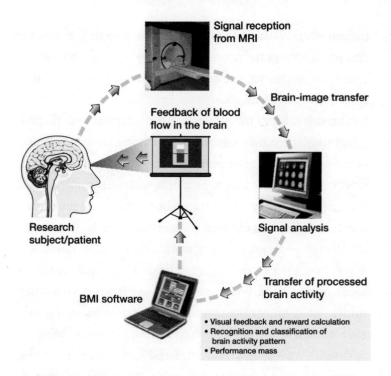

Configuration of a neurofeedback system for self-regulating brain metabolism

The person (left) lies in a magnetic-resonance scanner (MRI, top), which measures the blood flow in one or several areas of the brain. The person monitors her own brain activity, measured by detecting cerebral blood flow with functional magnetic-resonance imaging (fMRI), which is represented on a screen (centre) as a red or blue 'thermometer'. When the person increases the blood flow to a selected region of her brain, the thermometer shows red; when she reduces the blood flow, the thermometer's colour changes from red to blue. The computers (right and bottom) analyse the person's brain activity in real time so that she is always aware whether she has achieved the learning criteria.

The test subjects are given no specific information about how to influence the thermometer; they are simply instructed to observe when the thermometer reading increases, and the computer registers that the desired change in the brain has been reached. Just as they would when learning any new skill, subjects play around until they find the way to make the computer register success. Subjects then try to achieve this repeatedly, by intuitively (implicitly) triggering the same processes in their brain. How they achieve this, what thoughts or images they conjure up in their minds to increase the flow of blood to that part of their brain, is left entirely up to them. The system gives immediate feedback whenever the goal is successfully reached.

The 15 people who took part in our test needed only three neurofeedback sessions of four minutes each to learn how to significantly activate the front part of their insula. This was not just of theoretical value, as we were able to show in another experiment. People with an activated insula display much more intense emotional reactions to photos with negative depictions, such as mutilated war victims or abused children. This led us to expose antisocial psychopaths to our neurofeedback training, since it is known that they have underdeveloped insulae. The psychopaths found the training more difficult and required more sessions, but they eventually managed to increase blood flow to the insula region of their brains. And this also led to more intense emotional reactions when they were shown negative images.

This means the insula region can be changed, and in healthy subjects it doesn't even take very long. In our 'quest for emptiness', however, the insula would need to be deactivated, rather than activated, to shift our proprioception down a gear. A recent study we carried out with people suffering from an obsessive compulsion to wash shows that this is possible.

Using neurofeedback, we trained these patients to reduce the activity of their anterior insulae. In principle, this was done in the same way as training to achieve the opposite effect, only this time patients had to influence a visual symbol that represented a reduction in insular activity, rather than an increase. All our participants achieved some success in this during training, although to rather varying degrees. The participants managed to 'turn down' their anterior insula, and this effect was reflected in their behaviour. They showed far less shocked reactions to images of dirty people or objects. Their 'disgust threshold' was now considerably higher.

This means it should be possible to use neurofeedback to turn down the insula to such an extent that even the proprioceptive sensory complex is scaled back. However, neurofeedback is rather impractical for everyday use. There are now many clinics, psychology practices, and psychotherapists in Germany who work with neurofeedback, but they usually use EEGs, which are far less accurate in measuring blood flow in deeper regions of the brain than MRI scans. This is especially significant, because it is important to work on the correct part of the insula. Although we were able to modulate activity in the anterior part of the insula, proprioception takes place in the right posterior region of the insula, as described in the previous chapter. For these reasons, it has not yet been possible to prove that proprioception can be influenced at will.

In addition, it is necessary to consider what else can happen when the activity of the insula is altered. In Tübingen, we increased the activity in the insulae of patients with schizophrenia to enable them to recognise negative facial expressions. This was successful, but it also resulted in a decrease in their ability to recognise positive expressions. After training, they could judge whether someone was angry, but they were no longer able to tell when a person is happy. We must assume that the increased activity in the areas of their brains responsible for recognising negative emotions brought about a simultaneous decrease in the emotionally positive areas of their brains.

Reducing the activity of the insula to scale back proprioception could have a similar effect. The less responsive insulae of psychopaths result in not only less empathy, but also less fear than other people. If we reduce the blood flow to the insula, empathy, and consequently fear, may be reduced to such an extent that even healthy people develop psychopathic behaviours. We could end up with a subject who commits ruthless crimes in a state of profoundly relaxed emptiness.

All this makes it clear that neurofeedback is an interesting possible path to emptiness, because it does not require difficult psychological processes such as imagining or thinking, but there is much need for further research. For this reason, it makes more sense for us to concentrate on tried-and-true methods of achieving emptiness. One of those methods is meditation, especially Zen meditation, since it expressly seeks out emptiness.

Meditation is painful at first — but cannot function otherwise

When Janwillem van de Wetering arrived in Kyoto, Japan, he had big plans. His aim was to learn the daily meditation techniques of Zen monks in a real monastery. The Dutch author had saved enough money to see him through the three-year training if he lived frugally. He was determined 'to give up my previous existence and start a new life, which I could hardly imagine'.[2] He intended to leave the monastery only when he was overcome by complete enlightenment. But right from his first conversation with the master of the monastery, he had the uneasy feeling that he was on the wrong path. The old man told him in no uncertain terms: 'When you come to the end of the road and find perfect insight you will see that enlightenment is a joke.'

Life in the monastery was hard. Rising at three o'clock in the morning, going to bed at 11 in the evening. And in-between: gardening work and meditation, both painful. Before his first meditation session, Janwillem was told to adopt the lotus position: left foot on right thigh, right foot on left thigh, with a straight back, ears and shoulders in a line. He found it impossible, stretching his muscles too painfully. When Janwillem asked why he couldn't just meditate sitting in a chair, he was given the scornful reply: 'Why? Are you an old man? Or an invalid?' He was told that adopting the lotus position, in which the muscles are inactive but extended, is vital for the success of the exercise. This is because 'it is impossible to shape things if you are not in shape yourself'. Furthermore, he was told, you are less likely to fall asleep in the lotus position — and if you do, you are not so quick to topple over.

From our exploration of proprioception, we know that the receptors in stretched muscles send signals with that information to the brain, which responds by sending a message to relax the muscles. This has the effect of reducing muscle tone and calming the proprioceptive senses, which is a requirement for experiencing emptiness. Thus, it is likely that the lotus position creates the neurophysiological conditions necessary for achieving emptiness through meditation.

However, what Janwillem felt during his first meditation sessions was not emptiness, but pain. It was several weeks before he was able to spend any extended time in the lotus position. Then, his teacher gave him his first *kōan* — a kind of riddle for the meditating disciple to concentrate on. The important thing about these riddles is that they cannot be solved using logical reasoning, and not with any kind of target-oriented thinking.

A typical *kōan*, for example, goes: 'Everyone is familiar with the sound of two hands clapping. But what is the sound of one hand clapping?' Or: 'Show me the face you had before your parents were born. Show me your original face.' At first, most Zen disciples try to solve their given *kōan* rationally. But when they present the solution to their master, it is dismissed as unfit. The real meaning of a *kōan*, its main purpose, can only be revealed intuitively, without logical reflection.

The aim of the practice of pondering *kōan* is to recognise non-duality. The meditation practitioner aims to cast off the illusion that the universe is made up of individual things, and that the self has its own existence, distinct from the rest of the world. It is impossible not to think of Ian Waterman in this context, the man with no proprioception, who reported the feeling of having lost his self. It seems Zen meditation strives for that very experience of self-loss, and muting proprioception plays a key part in achieving that. The difference is that, unlike the case of Ian Waterman, it happens in a controlled way.

Meditation trains us to scale back our internal body perception to a point where we no longer feel any boundary between ourselves and the world around us. This is supported by the discovery by Japanese scientists of reduced activity in the posterior insula, where proprioceptive signals converge, in meditating practitioners. The activity of the thalamus — the gateway to the conscious mind — is also inhibited, while the anterior insula and the anterior cingulate cortex become significantly more active. These areas make up the 'salience network', which enables us to pick out from the constant barrage of sensory stimuli those which are most deserving of our attention.

The brains of meditators are especially able both to tune out the flood of stimuli they receive and to perceive intensely that which is important — without concentrating or focusing on any goal. However, 'important' does not mean personally important here, since that would simply be another kind of mind-wandering, far removed from emptiness. Another vital requirement for experiencing emptiness is that around the salience network there must be the extensive inhibitory effect on the brain produced by alpha waves, as described earlier.

But Janwillem had trouble with all this. It was mainly because he expected to gain things from meditation that he, having grown up in the hippie movement, believed could be expected of it. Things like enlightenment and transcendental consciousness.

After several months of intensive meditation practice, the author experienced a passionate moment of ecstasy. Suddenly, everything was beautiful to him: the goldfish in the pond, a pile of dog dirt, a dustbin. 'By losing myself in the colours and shapes around me, I seemed to become very detached,' he wrote, 'an experience I had known before after using hashish.' Only this 'monastery high' was far more satisfactory because he felt both happy and calm. The Zen disciple told his master about this experience — and was not even acknowledged with a silent nod. The master thought it was perfectly normal to find moss, dustbins, and goldfish visually interesting. A truism so obvious that it requires no further comment, particularly not from a Zen practitioner.

Rather than expecting to see progress from his meditation training, Janwillem should have been simply meditating and doing his work in the monastery, but he rarely managed that. Sitting in the lotus position was not only still painful, but also gave him haemorrhoids. Every session, he counted the minutes till his meditation torment was over. Occasionally, it was more bearable, and he felt as if there had been a change in his consciousness and he had thrown off the shackles of the self. This only ever happened when he sat down to meditate with the attitude that nothing would come of it and his efforts would be in vain.

It was not until later that the Dutch writer realised such a 'loser attitude' is an important driving force in Zen meditation. But, not having yet come to that realisation, he decided to lock himself in his room and meditate for three days and nights. He failed to reach that goal, and gave up the experiment after just half a day. And his stay at the monastery lasted one-and-a-half years, instead of three as planned. He came closer to emptiness, but was never able to reach it.

From sleeping in the lotus position to real meditation
I learned personally how far apart aspiration and reality can be among meditation practitioners while I was working at Munich University. We were using EEG to examine the processes in the brains of people while they were meditating. It must be said that these people were not practitioners of Zen meditation, but of the Transcendental Meditation techniques developed by Maharishi Mahesh Yogi. We asked them to give us a hand signal when they had reached a deep meditative state. Some managed to do that, but many did not — for the simple reason that they had fallen asleep. And even the EEGs of those who did give us a hand signal soon showed the slow wave patterns typical of sleep. They had done nothing other than that which we all do every night. The only difference was that they had managed to do it in the lotus position.

Later, in Tübingen, we examined the brain processes of

experienced Zen monks, and recorded very different results. The longer the monks meditated, the stronger the alpha and theta wave patterns in their EEG readings. Those are the activity patterns that usually occur while we are falling asleep, but not during sleep itself. It seems these Zen meditation practitioners were able to 'catch', i.e. prolong, the twilight waking state we feel as sleep descends upon us.

One of the main reasons for this was that their brains were in a state of heightened, but unfocused, alertness. For this to happen, all the senses — with the exception of proprioception — have to be working at full capacity. While meditating, these practitioners hear, see, taste, feel, and smell everything particularly intensely, but without focusing on or being interested in a particular object (known as 'absolute seeing, hearing, and cognition'). A Zen master once said, 'In the deepest meditative state, all impressions from the five senses are present, but do not trigger any inner thoughts.'

This is borne out by the results we collected using visualisation techniques to examine the brains of people meditating. These show that the posterior, perceptive areas of the brain are isolated from the anterior, executive areas during meditation. The sensory regions in the posterior part of the brain remain highly active, but their data is not captured by the anterior (prefrontal) part of the cerebral cortex. This means people who are meditating do attentively take in the world around them, but it is afforded no significance by them, and does not motivate them to action or control: things remain *empty* of meaning.

Above: a Zen monk meditating. Below: an EEG with dominant slow alpha and theta waves.

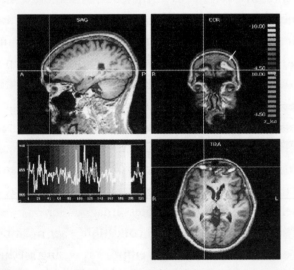

Functional magnetic-resonance image (fMRI) of a Zen master's brain while meditating: blue and green areas show that activity in the prefrontal regions of the brain (here, Brodmann area 10) is heavily reduced when the test subject indicates having reached the highest meditative state. Top left: lateral view; top right: frontal view; bottom right: transverse section, view of the brain from above.

Furthermore, the American neuroscientist Andrew Newberg recorded remarkable activity in the superior parietal lobes. The area stretches from the midsection of the brain to the back of the cerebral cortex. Newberg likes to refer to this as the 'orientation area' because it is involved with spatial orientation, and as such it creates a clear and constant perception of ourselves in space. In a similar way to the posterior insula, it helps us differentiate between self and non-self. As Newberg observed: during meditation, this region initially springs into action then suddenly deactivates when the full meditative state is reached.[3]

One possible explanation of this phenomenon is that the sensory centres at the back of the brain become detached not only from the meaning and control areas at the front, but also from the parietal lobes. The orientation area then receives no information, leaving two important functional units for self-perception — the posterior insula and the parietal lobes — high and dry. 'That leaves the brain with no option,' reasons Newberg, 'but to accept as real the perception that the self is infinite and closely connected to everything the mind is aware of.' The meaninglessness of all sensory impressions is joined by the disintegration of the self. Is any greater emptiness imaginable?

It seems there are certain conditions that must be met for a complete experience of emptiness during meditation to be possible. This explains why we were unable to record anything among the Maharishi adherents other than the fact that they had meditated themselves straight to sleep. The Zen monks from Asia, by contrast, were able to regulate their brain activity in such a way that they were neither asleep nor awake in the normal sense. This is why at first I believed their way of meditating was the more effective.

However, we then examined the brain activity of James Austin, a prominent proponent of Zen from the USA. He has been a practising Zen Buddhist since 1974, and is also a neurologist, so could be assumed to have no fear of our brain-scan tests. But when he signalled to us that he had reached a deep meditative state, his brain scan showed none of the activity described above. Separation of the anterior and posterior parts of the brain, disconnection of sensory perceptions from the areas responsible for meaning and self — he accomplished none of that. This was despite the fact that he had learned his meditation skills from the world famous Zen master Kobori Nanrei Sōhaku.

We can only speculate about why Austin was unable to produce the brain activity typical of the meditative state. In the same way as we can only speculate about what was going on inside the heads of the monks whose brains did produce those wave patterns. Despite all our modern visualisation technology, we are still unable to watch the brain thinking. The colourful pictures we get from MRI scans only show blood flow in the brain and the activation of and connections between individual areas, but they do not show actual thoughts or emotions. Yet if we add Janwillem's experience into the mix — the Dutch author's failure was primarily down to his inflated expectations — we can begin to get an idea about where Austin's problems might have lain.

As a neurologist, he certainly will not have been

intimidated by the MRI scanner, and so his defence system should have left him in peace in that respect. But he had spent a long time pursuing the disintegration of the self via Zen meditation and had carried out his own neurological research on the subject. It is reasonable to assume that he was anxious to 'show' us his skills. Unlike the Asian monks, who approached the MRI scan as they approach any situation — impartially and with no expectations — Austin probably had a goal in mind, namely to prove his own theories. And that is the worst possible circumstance for anyone attempting to achieve the senselessness and meaninglessness of emptiness.

Another study carried out by Newberg together with the psychiatrist Eugene d'Aquili shows that the attainment of an experience of emptiness is less dependent on the type of meditation practised than on the meditation practitioner and his or her attitudes. In their study, the two researchers monitored the brain activity of Christian nuns during prayer. As with the Zen monks, the activity in the nuns' parietal lobes first increased, and then decreased considerably during the prayer session. Only their subjective description of the experience was different. While Zen monks usually describe the meditation experience using allegorical concepts such as the dissolution of all distinct things including the distinct self, the nuns spoke of a feeling of being 'close to God' or 'merging' with him. However, these are all merely interpretations after the fact. The nuns' brain activity during prayer differed hardly at all from that of the Zen Buddhists during meditation.

What is the evolutionary significance of emptiness?

Followers of many other religions are also often called upon to abandon their 'self' as a way of becoming closer to the divine. Jewish mysticism, for example, deals expressly with the 'obliteration of the ego', and, three hundred years ago, Dov Baer demanded, 'Think of yourself as nothing, and totally forget yourself as you pray.' The mystics of the Greek Orthodox Church speak of *hesychia*, or inner silence, which opens the way to a feeling of unity with all things. Islam's mystics, the Sufis, developed the concept of *fana*, meaning passing away or annihilation (of the self). This was brought about by self-disciplinary and ascetic practices such as fasting, night vigils, and the endless chanting of verses from the Quran, to repress the self in order to make room for the divine.

The obliteration or emptying of the self is, then, one of the tenets of many religions. However, this is only to make room for the divine to enter into the place of the lost self. Only Zen Buddhism is willing to simply accept and explore emptiness per se, without immediately re-filling it with something godly. It may be a matter of contention whether this is the ultimate logical conclusion, or whether other religions have taken the concept further in viewing emptiness as an interim phase and a condition for divine revelation.

Whatever the case, emptiness can be described as a requirement for achieving the mystical religious experience of being connected to God, nature, or humanity. But that does not mean that this is where its *purpose* lies. That would be like saying that the reason the Sun exists is so that it can shine light on the Earth. The more pertinent question is why the brain developed and retained a love of emptiness during the process of human evolution.

The basic way evolution works is that adaptations that provide a concrete advantage for survival are favoured by natural selection. By the same token, mutations that are disadvantageous are selected against and disappear. The brain's inclination towards emptiness continues to this day, so it is likely to have been advantageous in some way for the survival of our species. But what concrete advantage for survival might it bring? When a prehistoric man was confronted with an angry cave bear, a tendency to zone out would be rather counterproductive from the point of view of the survival of the species. And if the caveman's self were obliterated, and with it the capacity for logical thought, then his ability to solve problems and develop strategies to fend off starvation or hostile tribes would also be extinguished.

Nietzsche was in no doubt about the meaning of emptiness for survival — without it we would suffer inner desolation. With its 'self-forgetfulness and transgression', emptiness is for Nietzsche the necessary 'Dionysian rapture', without whose burning power life would be 'corpse-coloured and ghostly'. That may be true from a psychological and philosophical point of view, but in evolutionary terms it is probably of only incidental importance. After all, there is no shortage of people in the world who survive and reproduce without ever experiencing the rapture of self-forgetfulness. A good night's sleep provides them with enough physical and mental regeneration. They may well lead 'corpse-coloured and ghostly' lives, but that is life nonetheless, and it knows how to perpetuate itself.

Newberg and d'Aquili have proposed a more plausible hypothesis for the importance of emptiness in human evolution. It postulates that our brain's propensity for emptiness developed in connection with mating and sexuality. Human beings are creatures with a very strong sense of self, which enables us to be aware of our own feelings, to assess them, and to modulate them if necessary. But for a species to survive through the generations, its members must reproduce, and for human beings this means emerging — for a short time at least — from this ego-cocoon we are in and merging with another person to form a single unit. The 'I' must temporarily become 'We', which is more likely to occur successfully if that temporary loss of self is experienced as something positive. And that is why, according to Newberg and d'Aquili, our brains love emptiness.

Newberg and d'Aquili go on to explain how it is 'no coincidence that the same language, with the same revealing terms, is used in religious mysticism and love'. Terms such as joy, delight, ecstasy, bliss, and devotion. In a certain sense, then, mystical religious experiences could be an accidental by-product of sexuality. It is the original feeling, which can also be achieved through meditation.

It must be noted, however, that as logical as this hypothesis sounds, it remains speculation. There is no way to prove that mystical experiences are a by-product of sexuality. But there can be no denying that sexuality and emptiness have much in common.

8

Lusting for Emptiness

*what sex, religion, and epilepsy
have in common*

Does sex make you unfree and stupid? Nietzsche suspected: 'The degree and type of a person's sexuality reaches up into the furthermost peaks of their spirit.' So we shouldn't set too much store by the independence of our spirit, since sex pulls its strings. Sigmund Freud adopted this idea for his layered model of the human mind, in which the ego is constantly torn between the moral superego and the instinct-driven id. This model is now considered outdated, but there is scientific evidence that our sexuality can drive us out of our mind — in the truest sense of the words.

Researchers at the Geneva University Hospitals used functional magnetic-resonance imaging (fMRI) to observe the brain activity of women with high and low libidos while the subjects were watching erotic films.[1] They found that the more sexually reserved women (low subjective levels of lust, infrequent intercourse) displayed a stronger activation of the frontal regions of the brain, which are responsible for attentive observation and higher mental functions such as self-control. In the test subjects with higher libidos, on the other hand, those regions became less active, while activity in the deeper 'lust centres' of the brain increased.

'You can never be wise and be in love at the same time,' Bob Dylan once said, and that's something students have been experiencing since time immemorial while trying to study for their exams: learning and lusting just don't go together. In fact, various studies show that lust can even lead to a radical drop in a person's IQ.

Unsurprising, because — as we saw in our discussion of mind-wandering — those whose thoughts keep returning to sex are neither in control of themselves nor in a position to come up with targeted solutions to problems. And in such a mind, there is no room for calm, discursive, objective, and disinterested assessment of the world.

We carried out a study to compare the brain activity of medical students who were 'head-over-heels in love' and those who were not in love — while they were calmly resting, while they were completing a cognitive task, and while they were thinking about their sweetheart (the subjects who were not in love were asked to think about their last romantic relationship). We found no difference when the subjects were calmly resting and while they were completing the task. As soon as the subjects thought about their old flame or their current crush, however, the complexity of the love-struck subjects' brainwave patterns suddenly took a dive. They became simple and predictable, and it seems safe to say the same was true of their thoughts.

LOVE IS BLIND

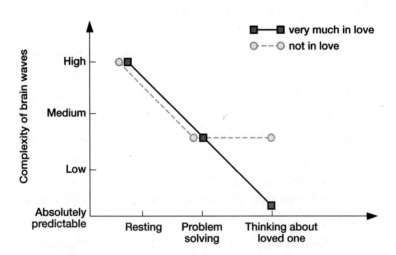

The illustration shows the electrical brain activity of students who are in love compared to those who are not. Both groups showed similar brainwave patterns while calmly resting and during concentrated problem-solving, but the brainwave patterns of those who were love-struck became far less complex when they thought about the object of their love. This is a clear indication of a decrease in networked thinking: the love-struck students' heads really were full of nothing but their loved one.

We could interpret the brainwave patterns of those love-struck students as the brain drifting towards emptiness, since it's clear that there is very little going on inside the heads of people in that state. But other studies have shown that people in love undergo similar hormonal changes and display similar behaviour patterns to people with compulsive disorders, and the continuous circling of thoughts around one particular thing is the very opposite of emptiness. So, in our search for a source of emptiness, we could dismiss sexuality out of hand — if it weren't for the sexual act itself, and its climax.

Toothbrush-triggered orgasms

You unscrew the top of the tube and squeeze a little toothpaste onto your brush. Then you open your mouth and move the brush in a circular motion over your teeth. You feel your saliva mixing with the toothpaste, producing a lubricated film that helps the brush complete its cleansing task more efficiently. Two minutes later, it's all over. You take the brush out of your mouth and rinse with a little water. Done.

Brushing our teeth is part of our daily hygiene routine, just like washing, blowing our nose, and wiping our bum. These activities are not normally very sexually arousing, and you need the imagination of a psychoanalyst to see anything sexual in the act of a toothbrush entering a mouth or in toothpaste-saliva lubricant. But for Emily, a young woman from Taiwan, that was not the case.

Emily was 22 when her womb had to be removed due to a serious medical condition. The operation went well; there were no complications. Except — the patient was subsequently unable to achieve orgasm, either during intercourse or by masturbating. Yet at some point, Emily's brain found another way to reach an orgasm.

At the age of 24, Emily suddenly felt sexually aroused while brushing her teeth, and very soon she was overcome by the muscle spasms typical of orgasm. At first, she dismissed it as some kind of 'mistake', but she continued to reach climax when she brushed her teeth — about twice a week on average! Sometimes, she would actually faint while orgasming, causing her to fall and injure herself. The young woman was plagued by feelings of shame, thinking herself perverse and obsessive, and so it was five years before she went to see a doctor.

Emily's doctor referred her to Chang Gung Memorial Hospital in Kaohsiung, Taiwan's second largest city, where her case attracted the interest of neurologists.[2] First, they wanted to find out what exactly was triggering Emily's orgasms. They stimulated the young woman's teeth and gums with chopsticks. Nothing, no reaction. Next, they had her smell and taste toothpaste. Again, nothing. They then asked the patient to move her hand as she did while brushing her teeth, thinking that perhaps the rhythmical motion was the trigger for her orgasms. Still, Emily showed no reaction. Only the complete toothbrushing procedure could make her climax.

When the researchers examined the young woman's brain using single-photon-emission computed tomography (SPECT) and magnetic-resonance imaging (MRI), they discovered an underdeveloped area in her right temporal lobe, and a clearly atrophied hippocampus. The medics also managed to make two EEG recordings of the woman's brain while she was orgasming during tooth cleaning. These showed the kind of spikes typical of the excessive neuron-firing during an epileptic seizure.

The scientists were now confident of the diagnosis: the woman was suffering from temporal-lobe epilepsy.

It has long been known to medical science that this can sometimes cause spontaneous orgasms. However, the problem in this case was that Emily's family had no history of epilepsy, and she had suffered no accidents, exposure to toxins, or diseases that might explain her epileptic seizures. Also, she was well beyond the age when temporal-lobe epilepsy usually develops — between the ages of five and ten.

Not to mention that there was still no explanation for the fact that she only ever had a seizure while brushing her teeth. Not when she drank alcohol, or while she was watching TV, dancing at a club, or driving; not when she was particularly frightened, nervous, happy, or sad. Only when she was engaged in the banal chore of brushing her teeth — an activity that hardly causes enough neuronal excitation to trigger a seizure.

Despite these uncertainties, Emily was prescribed the anti-epileptic drug carbamazepine. It did no good. So her doctors combined the drug with valproate, another anti-epileptic agent, because it works by suppressing neuronal firing along several pathways, for example by increasing the inhibitory effects of the neurotransmitter known as GABA. Under this medication, it took longer for Emily to climax while brushing her teeth, but that was the only change in her condition. This led the Taiwanese neurologist to state in her case report, 'Epilepsy and sex are similar in many ways.' They noted that this could also be seen in the fact that orgasms are accompanied by epileptic EEG patterns in the deep temporal lobes and the hippocampus.

As far as treating Emily went, the neurologists took a pragmatic approach, advising her to brush her teeth only briefly and rely on an antibacterial mouthwash. She now no longer orgasms during her oral hygiene routine. Unless, of course, she chooses to.

Sex frees you — of feelings, too

What exactly is going on inside our heads that makes sex and epilepsy so similar? To answer this question, we must take a closer look at the sexual act, from arousal via coitus or masturbation to orgasm.

The American physiologists Dean Dluzen and Victor Ramirez discovered a 90 per cent increase in dopamine levels in male rats when a sexually receptive female was placed in the cage with them.[3] An increase of that scale is a clear indication of desire, since dopamine is the transmitter used by the pleasure centres of the brain, such as the hypothalamus and the inner core of the nucleus accumbens. When the lab rats began to copulate, their dopamine levels spiked by another 10 per cent, and then levelled off again. This indicates clearly that dopamine and the brain regions with which it is interlocked essentially mediate the *desire for* an action, and not the *pleasure of* performing the act itself.

Human studies have confirmed this mechanism. For example, researchers measured a powerful but short-lived burst of activity in the dopaminergic areas of subjects' brains while they were watching an erotic film. Sexual stimuli leads to a powerful impulse of desire, which, however, levels off as soon as the object of that desire is attained. During the sexual act itself, dopamine's leading role continues to decrease.

Furthermore, it is not only the wanting that ebbs, but also sensory perception. During the desiring phase and during foreplay, our senses are heightened — the senses of smell and touch in particular become more acute than otherwise — but during the sexual act itself, we are fully focused on the matter at hand. The world around us fades into the background. So, from a sensory point of view, the sexual act can be compared with Peter Suedfeld's isolation tanks: we become immersed in a world of senselessness. This includes a reduced sensitivity to pain: many people perform acts of acrobatic prowess during sex that they would find excruciatingly painful under normal circumstances. Some people go temporarily blind, unable to see lights even when they are shining directly into their face, and many people lose the ability to judge their own physical capabilities. In 2015, a couple died in Mexico while having sex in a whirlpool tub when the overweight man collapsed and trapped his wife underwater.

A pivotal role in this 'zoning out' from the world around us is played by the thalamus. It closes its gates, putting consciousness on the back burner. This pushes us further and further towards a state of emptiness, eventually culminating in orgasm. As early as the 1980s, scientists discovered that during sexual climax, the right hemisphere is disconnected from the completely uninvolved left half of the brain, and alpha waves give way to theta waves, which usually appear when we feel sleepy.

In 2003, the Dutch anatomist Gert Holstege used

positron-emission tomography to measure the brain activity of men during ejaculation brought about through manual penile stimulation performed by the volunteer's partner.[4] This showed that while some parts of the dopaminergic system remained active, including the ventral tegmental area and the aforementioned nucleus accumbens, there was a general lull in the metabolic activity of the cerebral cortex. Men ride an extremely high wave of physical arousal during orgasm, as testified by a rapid pulse rate and frenzied breathing, but inside their heads the complete opposite is the case.

A few years later, Holstege managed to find 12 female volunteers for a similar experiment. They, too, were placed in a brain-scan machine while their partners attempted to bring them to orgasm by clitoral stimulation. Initially, this was unsuccessful, because the women kept getting cold feet in the chilly lab. However, after they were given socks to put on, they were able to climax. And their scans showed an even greater reduction in cerebral activity than the men in Holstege's previous experiment.

The left orbitofrontal cortex, which is responsible for controlling urges and self-control, was barely active at all, and the dorsomedial prefrontal cortex — responsible for self-control and social judgement — was also shown to be working in power-saving mode. This once again shows how important self-abandon and oblivion are for sexual fulfilment. What is perhaps more surprising, at first glance at least, is that important emotional centres of the limbic system, such as the amygdala, take a back seat during female orgasm, i.e. they become less active.

This led Holstege to conclude, 'At the moment of orgasm, women do not have any emotional feelings.' Their brains are so empty during orgasm, that they are no longer able to feel anything. Some women even pass out briefly — in the same way that Emily sometimes did while brushing her teeth.

Normally, we pursue something because it gives us positive feelings. Activity in the dopaminergic centres of the brain increases, and we are driven by feelings characteristic of wanting, such as desire, anticipation, urge, alertness, interest, lust, joy, and excitement. But, ironically, during orgasm, which is often described as 'the highest of all feelings', this mechanism is barely active, if at all. This is because it is dominated not by an abundance of happiness, but by the nothingness of emptiness. It is no coincidence that the French refer to orgasms as *la petite mort*, the little death. During orgasm, our consciousness becomes completely disconnected from everyday reality, up to the point of the loss of the self — and of our own history.

After interviewing 100 male and female subjects, the US researchers Paul Abramson and Steven Pinkerton came to the conclusion that the ability to remember feelings during the sexual act is woefully limited.[5] They also point out that most people also lose any sense of time; five minutes are 'remembered' as half an hour, and when individual experiences during copulation *are* reported more or less accurately, it is not due to concrete memories of them, but because most couples do basically the same things each time they have sex. One study among American emergency doctors revealed that annually five out of every 100,000 people suffer global transient amnesia due to orgasm — and this figure rises to 23 out 100,000 among over-50s.[6] Patients become disoriented and no longer know where they are or what is going on and tend anxiously to ask the same questions over and over again.

Despite this, there is clearly some appeal to this temporary loss of self and memory, since we are prepared to invest a great deal and take great risks to achieve it. At the extreme, no small number of couples try to increase the chance of attaining this experience of emptiness by means of oxygen deprivation. With nooses or neckties, with their partner's hands at their throat, with laughing gas, drugs, or plastic bags over their heads. The American Psychiatric Association estimates that two people in every million die during such 'rough sex play'. Autoerotic asphyxiation, in which people deprive themselves of oxygen while masturbating, can become a very self-destructive addiction.

Most people have little understanding for such sexual practices, as it is difficult to imagine someone putting their life at risk to achieve a more intense orgasm. Most parents of epileptic children are similarly horrified when their offspring wave their hands frantically up and down before their eyes to trigger a seizure. And when traditional South African sangoma healers dance around a cancer patient in a trance, most people will probably roll their eyes. But all these activities are about the same thing: creating emptiness, while ignoring the risk of the possible consequences.

Salvation and seizure: from epilepsy to religion

'For a few moments before the fit, I experience a feeling of happiness such as it is impossible to imagine in a normal state, and which other people have no idea of. I feel entirely in harmony with myself and the whole world and this feeling is so delightful that for a few seconds of such bliss one would gladly give up ten years of one's life, if not one's whole life.'

These lines were penned by the Russian writer Fyodor Dostoyevsky in a letter to a friend, who had great trouble believing them. Then, as now, most people saw epilepsy as a serious illness which would strike sufferers down, causing them to writhe on the floor and foam at the mouth, leaving anyone who witnessed it with 'a feeling of mysterious terror and dread', as Dostoyevsky himself once observed. This would seem to be light-years away from a feeling of blissful harmony.

Yet in the case of epilepsy, emptiness is often actively *sought* by sufferers. There are many online forums where epilepsy patients engage in a lively exchange about how to provoke a seizure. Methods include lowering blood sugar, staying up all night, drinking excessive amounts of coffee, or doing exhausting work. Running beneath the leafy canopy of a forest or along a garden fence can also provoke a seizure, due to the stroboscopic light effects this creates. Some users on those forums complain that their doctors have no sympathy with such behaviour. This isn't surprising, since an epileptic seizure always entails a risk of injury. Convulsive fits can also push many brain cells beyond their metabolic capacity, causing them to die as a result. Every seizure means the end for a significant number of functioning neurons. It was not for nothing that Dostoyevsky gave his highly autobiographical novel about Prince Myshkin the simple title *The Idiot*. Of course, many people with epilepsy are aware of these risks but try to provoke seizures nonetheless.

The reason for this is that an epileptic seizure follows a similar plot line to sex, culminating in something approaching emptiness. Its initial phase, known as the 'aura', can best be imagined as a build-up of more and more electrical charge in the neurons of the brain to the point where it becomes too much, and a massive discharge takes place in the form of a tremendous, overwhelming seizure. Just like an orgasm. Sexologists like to describe the phase immediately before male ejaculation as 'the point of no return'. The 'aura' that precedes an epileptic seizure functions in a similar way: in both cases, the person feels irresistibly drawn towards a state of emptiness.

When the neurological processes of epilepsy affect an area of the brain at the back of the left temporal lobe, the result can often be an intense religious experience. This was discovered by VS Ramachandran of the University of California after examining patients with temporal-lobe epilepsy.[7] The temporal lobes, which are the location of the neuronal thunder and lightning in this type of epilepsy, are closely linked to the hippocampus and amygdala, both anatomically and functionally. This localisation led the American neurologist to suspect that those with this type of epilepsy show a particularly intense reaction to any stimuli, like the Taiwanese woman who orgasmed while brushing her teeth. But Ramachandran's patients were focused on the spiritual: 'Their reaction to other categories, including sexual words and images, was unusually muted in comparison with other test subjects.' Instead, one patient reported, 'Suddenly, everything makes sense.' And, 'Finally, I have an insight into the true nature of the cosmos.' Others discovered 'the universe in a grain of sand', or felt they could 'hold the whole of eternity in their hand'. So, these people had mystical experiences in which the individual — including their own self — was superseded in favour of a greater whole.

This suggests that the divine enlightenment of many famous figures in the history of religion was due to temporal-lobe epilepsy. Such visionaries as Moses, Ezekiel, Mohammed, Saint Teresa of Ávila, or Saint Paul, who famously had a vision while on the road to Damascus, in which Jesus appeared in a great light and spoke to him. As a child, Ellen White, the American founder of the Seventh Day Adventist Church, was hit on the head by a stone thrown by another child, which resulted in repeated seizures accompanied by visions; and Dostoyevsky believed he could touch God during his fits. George Fox did not have any visions, but he was such a vivid and convincing preacher that his followers would quake with paroxysms of religious ecstasy, hence the commonly used name of the movement, the Quakers, although the group actually calls itself the Religious Society of Friends. In Siberia and some parts of Africa, shamans still dance themselves into a trance-state to the sound of rapid drumming and the shaking of rattles, with the aim of exploring other worlds on a 'spirit journey'.

However, when an individual's personality is surrendered in favour of the divine, that person can lose sight of others' lives and wellbeing. In a survey carried out among Canadian university students, 7 per cent of those questioned said they would be prepared to kill if God told them to do so. Among male students who regularly attended a weekly religious service and had also experienced religious revelations and epileptic seizures, the proportion of potential killers rose to more than 44 per cent. The lead scientist in the study, Michael Persinger of Laurentian University in Sudbury, Ontario, imagined that number would probably be even higher 'if responders were recruited directly at a church rather than a university'.[8]

For most of us, religious fanaticism is not an option for achieving emptiness, and the majority of healthy people see epilepsy as something far removed from themselves. They are presumably more likely to see themselves using orgasms as an effective method. But is it really a way to enter emptiness for any appreciable period of time? US researchers estimate the average duration of the male climax to be 12.2 seconds, which means men spend a total of only just over nine hours of their lives at the peak of sexual lust. That is not very much in view of the fact that life expectancy has now reached more than 80 years. Moreover, as we have already seen, men's brains are not as completely overtaken by emptiness during orgasm as are women's. Overall, however, women experience orgasm less frequently than men. One study involving 4,000 women found only 14 per cent regularly achieved orgasm during sexual contact; 32 per cent of respondents said they reach climax only a quarter of the time.[9]

So there is good reason to consider other sensual strategies for achieving emptiness. And epilepsy is interesting in this context, since it is not quite as far removed from us as we think.

With Pokémon into emptiness

Children's television is a tricky business. It is impossible to predict what the next popular fad among kids will be. But in 1997, the Japanese broadcaster TV Tokyo landed a hit with its series *Pokémon*, which had nearly every child between the ages nine and 12 glued to the screen. Although the cartoon featured monsters and baddies, the plot lines were pretty harmless, so parents and educationalists saw no cause for concern.

In December 1997, an episode with the title *'Dennō Senshi Porygon'* ('Cyber Soldier Porygon') aired. The plot was innocuous enough, but the episode included a scene in which the Pokémon character Pikachu destroys two missiles with his 'thunderbolt', causing an explosion, which was represented on screen by flashes of red and blue lights. The whole explosion sequence lasted for little more than six seconds — but this was enough to trigger epileptic seizures in several thousand Japanese children. Just under 800 were taken to hospital by their worried parents. There were no lasting negative effects, but the general public were understandably alarmed. The news media pounced on the story, and some replayed the offending sequence in their reports, causing seizures in several hundred more children.

What had happened? It later turned out that none of the children affected, at least none of those who were hospitalised, had a history of epilepsy. Nonetheless, they had suffered seizures. Some had convulsions; others froze in front of their TV sets, staring wide-eyed at the screen — a phenomenon known as an 'absence seizure'. The reason for these reactions was that the children's brains had picked up and adapted to the stroboscopic rhythm of the light flashes on the TV screen and switched to a theta rhythm, which occurs not only during orgasm, but also during the phase that scientists call the 'twilight state', somewhere between sleeping and waking. You could say the children had been 'beamed' into the same kind of emptiness we experience when we climax sexually or when we lie dozing in a hammock. Neither of those states is generally thought of as unpleasant, which explains why, rather than subsequently avoiding television, many of the children affected would actively seek it out, sitting expectantly in front of the TV hoping for a repeat of the *Dennō-Senshi-Porygon* effect. Not surprisingly, their parents were concerned.

The Japanese phenomenon, which became known as the Pokémon Shock, is an indication that healthy individuals should not necessarily reject epileptic states out of hand. Adult brains are affected less often than children's, but anyone who thinks they are immune to the effect need only watch the faces of the people on the dancefloor in a club when the strobe lights are turned on in addition to the rhythmic beat of the music. Dancing in these circumstances is unlikely to make you lose consciousness or freeze on the spot, but the chances are high that a strange kind of stillness will enter your perceptive world and your inner world of thoughts. Our brain is a resonator that loves to 'get into the groove' of any rhythmical oscillations in its environment.

9

The Rhythm of Emptiness

how music carries us away

'It's gotta groove. The lead guitarist doesn't need to be a speed freak, and the bassist doesn't have to be a fine technician. But when we all get into the groove together, everything is alright. And the audience can feel it, too. They're with us. The whole room is in the groove. And moments like that are the reason we make music.'

Andy has a clear idea of what makes a good gig. The 56-year-old from Bremen has been playing the drums since he was a teenager, mainly in jazz and blues bands, and has had many different experiences of gigging over the years. He's played with his bands to packed venues for good money, without getting any kind of 'buzz' out of it. A nice fat pay check was a reason to be happy, naturally, and it is always pleasing to know you can fill an auditorium — 'but none of us got really into the music'.

By contrast, Andy has also played to half-full halls, and nonetheless there it was: that feeling of getting 'into the music'. The band rocked like there was no tomorrow, and the few audience members were swept along. Feet tapping, hands clapping rhythmically against thighs, heads nodding or swaying in time, dancing. Everything was moving. 'Maybe it was because we just went out and played, with no pressure,' Andy conjectures. After all, there were not enough audience members present to hurt the band's pride with criticism or lack of interest.

But, of course, Andy says, the best times are when you have 'a combination of great music, a packed venue, and a pumped crowd'. That feeling is indescribable. And he means this literally, as in 'not able to be described'. 'I don't have the words,' he says, and adds that he often can't remember any details about gigs like that. 'But afterwards you just feel a fantastic buzz.'

Baby groove: even newborn babies move to the beat

A group all getting into the groove together — it's not only musicians who are familiar with this. Most music-lovers and dancers know the feeling, too, even if they can't explain the meaning of the term, which first became popular in the second half of the 20th century and originally goes back to the groove, or furrow, that guided medieval ploughmen across the field. They start tapping their feet as soon as Benny Goodman's 'Sing, Sing, Sing' strikes up or Beethoven's Seventh (described by Richard Wagner as 'the apotheosis of the dance') comes on the radio. Or when they start jumping up and down at a rock concert as the band kicks off in four-four time.

This propensity to groove on down with the music is one of the main things that separates humans from animals. Wolves howl, cats yowl, and birds and humpback whales even sing tunefully. But when you see a polar bear at the zoo rocking its upper body rhythmically to and fro, it has less to do with music than with a pathological condition that would be called hospitalism in human children. Grooving along to music is a phenomenon unique to human beings. And it seems we are born with this ability, as was demonstrated impressively by two studies carried out in recent years.

In one of those experiments, the Hungarian psychologist István Winkler played a rock-style composition, with a strong drum and bass line, to 14 newborn babies while they were sleeping.[1] EEG readings were used to measure the babies' reaction to the music. The piece began with a clearly structured rhythmic pattern, but, as it proceeded, individual drum beats were left out, which caused significant changes in the babies' brain activity — particularly when the missing pulse was on the downbeat. 'The babies appeared to be so into the rhythm that they felt disturbed when the principle beat for that rhythm was removed,' Winkler explains. While the beat pattern remained unchanged, there was also no reason for the babies' brains to change mode. But the irregular beat caused a reaction — and the neurons started firing.

In the second experiment, psychologists Marcel Zentner of the University of York in England and Tuomas

Eerola of Finland's Jyväskylä University analysed the way babies aged six to 24 months move to music — and what effect those movements have.[2] The children were seated on their parents' laps in such a way that they could move their heads, arms, and legs freely. They were then played various pieces of music, including Mozart's *'Eine kleine Nachtmusik'* and Saint-Saëns' *The Carnival of the Animals*, as well as interpretations of the same pieces reduced down to little more than their rhythm, and also children's songs and rhythmically very varied drum beats. The children were also played a recording of an adult voice reading aloud as a non-musical control stimulus. The infants' reactions were recorded on video.

The results showed that babies moved rhythmically only when listening to music, and that their movements were more frequent and more pronounced the more rhythmical the music was. It should be noted that none of the children came from households where music was part of their daily lives. The children's reactions did not vary with their age: the test subjects who were only five months old moved to the music in exactly the same way as those who were older.

The movements of half the children were analysed by eight professional dancers, who judged how well the babies' movements were synchronised with the beat of the music they were hearing. The more positive the dancers' assessment was, the more often the babies were seen to smile or laugh: the children clearly felt pleasure when they got into the groove. This led Zentner and Eerola to conclude: 'The findings are suggestive of a predisposition for rhythmic movement in response to music and other metrically regular sounds.'

And we would add to that conclusion that the background to this predisposition is the achievement of emptiness. The evidence for this has already been presented. Many of the activities that can lead to emptiness are connected with rhythmical movement. We need only think of masturbation and sex; the 'merging of bodies to form one rhythmically pulsating unit' is one of the noble objectives of sex therapy. When we are swinging gently back and forth in a hammock, we feel our mind winding down and a pleasant sense of relaxation overcomes us — a phenomenon that has now been proven scientifically. And when we 'drift off' during a train or ship journey, it has much to do with rhythm: the rise and fall of the waves or the rattle of the tracks. It is no coincidence that every fun fair includes rides that shake us about, not only violently, but also rhythmically.

Meditation usually also includes a rhythmical element, such as measured breathing or the chanting of mantras. In religions such as Candomblé, Voodoo, or Sufism, for example, euphoric dancing is used to attain a state of religious ecstasy. When the Whirling Dervishes of the Mevlevi Order spin around at great speed, it may look like a nice dance performance to us, but for the Mevlevi themselves it is a devotional act that helps them transcend the worldly sphere and come closer to God.

The Jesus Prayer advocated in the Orthodox Christian churches is very tranquil by comparison, but it also relies heavily on rhythmical elements for its effect. Believers repeat the name of Jesus Christ over and over again until their chant becomes synchronised with their breathing pattern or heartbeat. By this time, they are usually in a state of rapture and oblivion, still able to hear their own prayer, but no longer able to influence it by their will.

Jews and Muslims rock their upper bodies back and forth as part of their ritual prayers, and, when Catholics pray the rosary, they are often transported by the rhythm of the prayers and 'mysteries' they recite for each of the 59 beads.

No wonder rhythmical music can help us approach a state of emptiness.

Nourishment for the brain and for emptiness

Neuroscientists have long been intrigued by the phenomena of music and music-making. This is partly because playing a musical instrument is top-quality mind food for the plastic brain, since it affects not only the centres of the brain that process acoustic signals, but many other parts of the brain besides. For example, the motor regions are activated to move the fingers of the pianist or guitarist, or the facial and respiratory muscles of a wind-instrument player. The areas responsible for touch are active, since it is necessary to have a feel for an instrument, in the truest sense of the word, in order to play it. And the visual centres are vital in reading sheet music or watching a conductor or fellow musicians and much more.

In Tübingen, we were able to show that music also improves proprioception, that is, our sense of our own body and its movements. On closer inspection, this is not surprising, since a pianist, for instance, needs to be able to feel how forcefully and for how long she presses each key, or her playing will be devoid of dynamics. And of feeling, which brings us to another central brain function: good musicians must be emotionally involved in the music they produce — and that means activating the corresponding brain regions.

Furthermore, all these actions do not take place sequentially, but more or less simultaneously, which requires not only physical coordination, but also associative skills. Thus, actively playing a musical instrument challenges and trains virtually the entire brain in multiple ways. This helps improve, or at least stave off, deteriorations in the functions of the brain, which explains why active musicians are apt to develop dementia comparatively late in life. Yet all this has nothing to do with emptiness, since a brain being challenged in multiple ways is the precise opposite of empty.

However, as we know, there are many different genres of music, and musicians can differ greatly in their skill levels. As we were able to show in Tübingen, both of these issues have a major influence on whether music can create a pleasant emptiness.

HOW MUSIC INFLUENCES THINKING

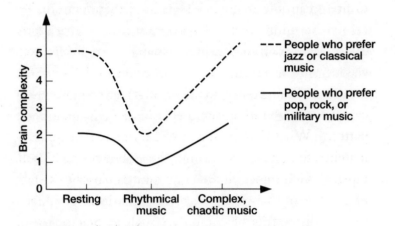

The illustration shows the complexity of the neuro-electrical processes recorded in the brains of people listening to more rhythmical pop, military, or folk music on the one hand, and those listening to more melodic classical music or jazz on the other.

We calculated the complexity of thoughts using mathematical algorithms taken from the field of non-linear dynamics (better known as 'chaos theory'). This involves calculating the frequency with which a particular brainwave pattern is repeated during completion of a task, and therefore how predictable that pattern is. A high level of complexity equates to a low level of predictability of the brainwave patterns. A low level of complexity, on the other hand, indicates a simple structure and high predictability of brainwaves — and therefore a closer proximity to the brainwave patterns typical of emptiness.

The illustration shows that jazz fans and classical music lovers generally develop more complex thinking patterns, which initially take them further away from a state of emptiness than those who prefer rock and pop music. However, we also see that complexity significantly reduced in both groups when they listen to highly rhythmical music with a simple melody.

We wanted to examine the different ways listening to different genres of music would affect the electrical brain activity of those listeners. Our results showed that simple

melodies in which rhythm is dominant cause neurons
to fire off in low-frequency alpha and theta patterns —
across the entire cerebral cortex in equal distribution.
Those are the dominant patterns when we are in a relaxed
waking state or during the dozing phase shortly before we
fall asleep. Brain activity generally displayed a low level of
complexity with rather stereotypical and predictable wave
patterns. What that means in real terms is that military
marches, folk music, pop, samba, blues, boogie, rock'n'roll,
hip-hop, Goa trance, and techno take us closer to a state
of emptiness than classical music, free jazz, or indeed
the serialist works of composers such as Schönberg or
Stockhausen.

Among professional musicians, we even found brain
activity similar to that of meditators. On the one hand,
extensive, low-frequency theta and alpha waves; on the
other, small islands of electrical hyperactivity projecting
out of this gently lapping sea, indicating the musicians'
high level of attentiveness. We discussed this mechanism
in detail in the context of the 'salience network', in
Chapter 7.

This indicates that the brains of excellent musicians
(like those of meditating Zen monks) are especially
able to cut themselves off from the flood of stimuli and
let their 'thought pump' run idle, while simultaneously
improving their perception of that which is important
— without concentrating or acting in a target-oriented
way. In this case, 'important' means chiefly that the music
is perceived and played *with emotion*. Schopenhauer
would say that when that happens, the will can see itself

in the music (see Chapter 2). And this works so well in professional musicians because their physical mastery of their instruments is so highly developed that they are able to play without the need to set complex brain activity in motion. The necessary physical sequences have become so automatic and anchored in the deeper parts of the brain, such as the cerebellum, in these musicians that they hardly need to make use of their cerebral capacities at all.

By contrast, it is rare for amateur musicians to reach this state, because they need to concentrate on the actual playing of their instruments. The sensory and motor regions of their brains are then especially active. This means their brains cannot produce a uniform sea of low-frequency waves out of which they can raise the salience network's 'rock of attention'. Amateur musicians do not experience sufficient emptiness, leaving them lacking the clear view that is necessary to feel the essence of the music and communicate it to their listeners. This is why we are more likely to feel moved when a piece is performed by an experienced soloist or orchestra, whether the music in question is 'Hoochie Coochie Man' or Beethoven's Fifth. The important factor is whether the musicians' playing is good enough to produce sufficiently regular brainwaves — viz. emptiness — in their heads while performing.

Better timing with the bass

First-rate musicians, then, achieve a state of emptiness more easily than someone with little experience of music-making. However, simple, highly rhythmical sounds can take us deeper into emptiness than music with more complex structures, which require the brain to engage in similarly complex thinking in order to process them.

It is no coincidence that many people reach a state of rapture when they listen to military marches, and the irresistible rhythms of samba can lead to euphoria even outside of the carnival season in Rio. In the 1930s, Benny Goodman's laid-back swing music sent the audience crazy at Carnegie Hall, which was usually associated with serious and solemn classical concerts; and when Bill Hailey, Chuck Berry, and Elvis Presley rocked the house 20 years later, the audience would sometimes wreck the hall furniture. These days, ravers dance themselves into a state of ecstasy to the beats of techno music or hip-hop, and, even though rap is ultimately about the words, they are carried by the heavy groove of a computer beatbox. Wherever we look in music history, we see that ecstatic oblivion is associated with a heavy, rhythmical beat.

This raises the question of whether slow or fast rhythms are more likely to set the brain resonating. As a reminder: the theta and alpha waves responsible for alert wakefulness and the 'twilight state' have frequencies between 3 and 13 Hz. Frequencies like that cannot be reached even by speed metal or rockabilly music, let alone slow blues as played by someone like BB King. But the point is not to find a one-to-one match for the beat of the music and the frequency of the brain's waves; the point is the groove, and that is not the same as pure beat.

When a slow-blues drummer plays quavers on his hi-hat, he will certainly match the frequency of theta-wave activity, and, by getting into that groove as a listener, my brainwaves will willingly synchronise. Even if those quavers are not played explicitly, my brain may still be counting them. Beat is an objective unit in music — but groove is a subjective pattern, which musicians and listeners get into together. Or in the words of Andy, the drummer from Bremen, 'You can play incredibly fast but still be in a slow groove; and you can play incredibly slow and be in a fast groove — what matters is that everybody agrees on *how* they experience the rhythm. If there's no common denominator, there can be no groove.'

That common denominator can only become established, however, if there is an impulse from below, this is, from the bass. Canadian researchers discovered that the bassline helps us keep in time and thus creates the foundation for a groove. The team of scientists led by Laurel Trainor of McMaster University in Hamilton, Canada, had 17 test subjects listen to two simultaneous piano tones, one higher and one lower, which were repeated every half a second. That timing was at times minimally varied; sometimes the onset of the higher tone was delayed by 50 milliseconds, and sometimes it was the lower tone that was delayed. During the experiment, the researchers used EEG to measure the subjects' brain activity. They found that the brain's reaction to rhythmical deviation in the low tone was much more pronounced than in the higher tone.

In a second experiment, the subjects were asked to tap along in time to the notes, which were once again staggered by 50 milliseconds. This only confused the participants when it was the bass note that was delayed. When the higher note was delayed, the subjects were more or less able to keep time, which was taken by Trainor as a clear indication that 'the lower tone has greater influence than the high tone on determining the perception of timing'.[3] This bass impulse need not necessarily come from a tuned instrument such as a piano, double bass, cello, or tuba, but can also come from a bass drum. But if that impulse is missing, timing is quickly lost and with it the chance to get into the groove. We can only nod along when the bass is booming.

And our heads nod more accurately in time to the groove if the music we are listening to is from a familiar cultural background. A Canadian study involving African and North American subjects showed that while both groups were able to perceive and differentiate the rhythms of the other's culture, they had great trouble tapping their hand along to the foreign beat, and, correspondingly, their brain activity was dominated by high-frequency wave patterns.[4] Low-frequency waves were only present to any significant degree when the subjects were tapping along to the familiar rhythms of their own culture. In considering this study, however, it must be remembered that the North American subjects were not completely unfamiliar with the African groove, since much of the music in their own culture has its roots there. It hardly bears thinking about how these results might turn out if a Ruhr-valley steelworker or a Viennese confectioner were asked to tap along to the rhythms of Brazilian samba or Japanese taiko drummers.

Adele, Rachmaninov, and 'skin orgasms'

But even if a rhythm has a more powerful effect on us the more familiar it is, it does not follow that music always has to be the same in order to take us to a state of emptiness. Quite the contrary, in fact.

Psyche Loui of Wesleyan University in Connecticut has spent much time studying the criteria a piece of music must fulfil to elicit such a strong emotional response that we are transported to a state of ecstasy. Loui herself first experienced this as a student, when she heard Rachmaninov's Piano Concerto No. 2. 'I was instantly captivated,' says the neuroscientist. 'A chill down the spine, fluttering in the stomach, a racing heart.' In fact, not very different from an orgasm. The feeling was so overwhelming that Loui decided to focus her work on this phenomenon — referred to by scientists somewhat misleadingly as a 'skin orgasm'.[5] She sifted through the academic literature and carried out her own experiments, using EEG to measure the brain activity of test subjects while they listened to music.

Her work confirmed the finding that music produces simple, low-frequency brainwave patterns via rhythm. However, the ecstatic state of emptiness only occurs when the listener's expectations are violated. 'This can be a sudden change in loudness, for example,' explains Loui, 'Or a sudden change of key, a spontaneous phrasing of the melody, or a syncope in the rhythm.' All of those cause the 'rock of attention' described earlier, with its associated absolute alertness, suddenly to rise out of the sea of theta and alpha waves — and the result is a state of ecstasy.

Rhythm carries us along towards emptiness, but we reach the destination by having our expectations of conventional music violated. Yet that break must not be so violent that we experience it as disturbing or even frightening. Rather, we need to experience it as sensual. We temporarily lose control, and that loss excites us. Like the ice-cold plunge pool after a sauna, which, rather than shock-freezing us solid, gives us a pleasant shiver down the spine. Since we know that taking the plunge is not going to kill us.

The American blues performer BB King was a master of the art of sending people into an exceptional state of ecstasy with his music. During his live concerts, he only rarely strayed from the typical chords and shuffle rhythms of blues music, allowing the brains of his audience easily to slip into that groove. But King's singing was extremely varied, switching suddenly from powerful shouting to a sensitive falsetto, or he would draw out the intro before he began singing for so long that some audience members thought he had missed his cue. As a guitarist, he loved extended notes that contrasted with the constant shuffle beat built up by his band, and he was able to turn the volume right down, only to suddenly blast the audience away with blaring horns. His breaks were legendary, and, when they ended and the band fell silent, everyone in the audience would be left still feeling the groove. In that emptiness, nothing could be heard — apart from the wails and ecstatic screams of a few fans.

Those fans had known what was coming, and yet they were still taken by surprise every time it did.

Because they had willingly let themselves be drawn in by the uniformity of the blues groove, they were able to experience the breakdown of that structure as a sensual descent into emptiness. Which once more shows that one of the prerequisites for emptiness is a willingness to set out wholeheartedly on the path to achieving it.

Psyche Loui has drawn up a list of pieces that fulfil the criteria described above for provoking an ecstatic reaction particularly well.[6] Alongside a toccata by Bach, it also includes the song 'Someone Like You' by Adele, and 'Wonderwall' by Britpop band Oasis. And, of course, Rachmaninov's Piano Concerto No. 2, of which Loui says, the 'harmonic twists in the second half always get me!' Although, like those BB King fans, she is familiar with every note. On the path towards emptiness, there need be no fear of repetition.

Even more important than that, though, is that there should be no fear of losing something in emptiness. This is also the case with many types of mental illness, as we will see in the next chapter, including depression and borderline personality disorder, schizophrenia and dementia, ADHD and psychopathy, which could well be called 'diseases of emptiness'. However, we will also see that many of the problems associated with those conditions do not spring so much from emptiness itself, but rather from the fact that sufferers find leaving behind their previous effect-based life so difficult, or it is made difficult for them by other, possibly well-meaning, people.

For example, dementia patients who are constantly made to complete endless crossword puzzles or so-called brain-jogging exercises will become increasingly despairing in the realisation that they cannot halt their memory loss with such exercises, and the repeated lack of success will only serve to underscore the relentlessness of their progressive loss. Yet if they can accept the invading emptiness in their head, it can become a kind of life jacket for them.

10

The Pathology of Emptiness

how we should deal with 'diseases of emptiness'

In its recent history, psychiatry, like other branches of medicine, has seen a constant increase in the number of conditions regarded as requiring treatment. The list now includes around 100 disorders, from Alzheimer's and depression to schizophrenia and obsessive-compulsive disorder. One reason for this huge number is a vigorous process of diversification: creating new medical conditions opens up new job and earnings opportunities for doctors and therapists. Numerous studies have shown that the willingness among medics to diagnose an 'illness' increases if they believe there is an effective medication to treat it. On the other hand, the long list of conditions is also a result of efforts to bring some order to the chaos of abnormalities and disorders of the brain. Obviously, the pathology of each condition must be described in detail, otherwise there would be no way to distinguish efficiently between them.

The problem with this is that focusing on details often leads to a failure to see the features that certain conditions have in common. And in the field of psychiatric disorders, these commonalities are likely to be rather numerous since they all share a common denominator: the brain. This organ is extremely plastic, allowing it to adapt very well to the conditions of its environment, and allowing for a great variety of possible disorders. But the brain also has a special relation with that which always stays the same for the simple reason that it is, literally, nothing: emptiness. As we have seen, that relationship can be extremely fruitful and lead to enlightening insights into our own existence. But it can also be extremely problematic and associated with fear — and as such, it can be a driving force for many disorders.

Psychopaths, for example, feel repeatedly drawn to extreme behaviours by the feeling of emptiness. One of their most important characteristics is their boredom-driven (emptiness-driven) search for sensation. And which person with depression is not familiar with the devastating feeling of being trapped in the vacuum of futility? Such mental disorders are the 'diseases of emptiness'.

As if every cell were as sick as a nauseated stomach: the black hole of depression

'Well this … isn't a state. This is a feeling. I feel it all over. In my arms and legs … All over. My head, throat, butt. In my stomach. It's all over everywhere. I don't know what I could call it. It's like I can't get enough outside it to call it anything. It's like horror more than sadness. It's more like horror. It's like something horrible is about to happen, the most horrible thing you can imagine — no, worse than you can imagine because there's the feeling that there's something you have to do right away to stop it but you don't know what it is you have to do, and then it's happening, too, the whole horrible time, it's about to happen and also it's happening, all at the same time.'

This is how Katherine describes her depression, or rather, attempts to describe it. Few things are more difficult than expressing this mental disorder in words. It is not, as many people assume, just a particularly severe type of sadness; it is not a reaction to something in real life making a person sad. Rather, depression is more like an ill-defined inkling of something unimaginably terrible, and, at the same time, a feeling that the consequences of that unimaginable horror are already at work, even though the dreaded thing has not become reality. This is difficult to imagine for someone in robust mental health, since they are used to considering events in a rational time sequence. They do something, then they ascertain whether it was successful or not and adjust their future course of action according to those results. But people with depression see a futility in all actions even before they are taken, and therefore end up doing — nothing. 'What's the point of washing,' asks Katherine, 'if everything smells like I need another shower?'

The 21-year-old data-entry clerk is a character in the novel *Infinite Jest* by the US author David Foster Wallace.[1] We must assume that she shares many characteristics with her creator, as Wallace, formerly a talented tennis player and an eloquent superbrain among the writers of his generation, had been through the 'full program' typical of a depressive's career: alcohol, cigarettes, electroconvulsive therapy, and, of course, lots of medication, which he repeatedly stopped taking because he wanted to be able to function without it. His depression made him feel like someone who had been poisoned, as if 'every single cell in your body is as sick as [a] nauseated stomach'. Eventually, he saw only one escape from his 'one-man hell': on 12 September 2008, he hanged himself at his home in Claremont, California. He was 46.

There is scarcely another writer able to describe the hopeless, senseless vacuum of depression as powerfully as David Foster Wallace. Sometimes he speaks of being submerged in an all-encompassing body of water and then, struggling for breath, realising that there is no surface and, in fact, no beyond. Another time, he quotes an image from the famous novel by his fellow depressive and suicidal author Sylvia Plath: that of the bell jar, separating him from the rest of humanity, with the oxygen inside slowly running out. And finally, he compares depression to a black hole, swallowing up all of existence, and even slowing down time, until it eventually comes to a stop, like a river robbed of its gradient. Wallace was constantly concerned with the idea of depression as a vacuum of hopelessness and senselessness, in which even the logic of time eventually seems to get lost, and that idea coloured his creative output. Interestingly, science now supports this model.

A group of researchers led by Jim Lagopoulos at the University of Sydney compared the brains scans of more than 1,700 severely depressed patients with those of some 7,200 people with no mental health issues — and found significant differences. On average, the depression patients' hippocampi were 14 per cent smaller than those of the healthy controls. This difference was significantly influenced by the number of depressive periods from childhood. For Lagopoulos, this is a clear indication that it is the disease that changes the brain.[2]

Severe depression usually occurs after an extended period of absolute helplessness: the death or other loss of an important person (e.g. parents' divorce, child leaving home), lengthy incarceration, torture, the aftermath of natural disasters, and war. The long-term effects of such experiences often include post-traumatic stress syndrome (PTSD), in which the events that triggered the original trauma reappear as 'flashbacks' — as also seen in drug addicts — for example as shocking images during sleep. Overall, however, it is the — mostly illogical — thoughts about the consequences of their helplessness that are at the forefront of the minds of people with depression: 'The fact that this has happened proves that my life no longer has any meaning.'

The experience of helplessness floods the brain with cortisol and other stress hormones, and this affects the hypothalamus and the hippocampus. One result is that fewer neurotransmitters such as NGF (nerve growth factor) become active in those areas. These substances are necessary for the construction of synaptic connections.

Levels of adenosine, which nerve cells require to produce energy, can also fall drastically. This significantly reduces the ability of both brain regions to function properly, and the eventual result is that depression patients sink further and further into an all-encompassing emptiness of mind.

Deep within the brain, the hippocampus ties together individual memories in the cerebrum to create a comprehensible whole that we can understand. It not only tells us that the bulges on the edges and in the middle of a person's face are their ears and lips, but also lets us know that they belong to our mother or lover. It not only tells us that a door is a door, but also lets us know that we can use it to enter or leave a room. It is the hippocampus that makes sense of our individual sensory perceptions by binding together the flowers of our memory into a bouquet. This mostly occurs during deep sleep, which is initiated by the hypothalamus in the first three hours of the night.

The memory-tying process is suppressed in people with depression, not least because of the sleep problems that many struggle with. Their hippocampus, as the memory's 'giver of meaning', becomes stunted, which goes a long way towards explaining the inner emptiness felt by these patients. Their lives become meaningless, in the literal sense of the word. The people around them, their work, food, sex, any activity loses all meaning. Why wash, why speak, why work, why even breathe when everything is meaningless?

The Danish philosopher, Søren Kierkegaard wrote, 'Life can only be understood backwards; but it must be lived forwards.' For people with depression, both lose their validity: they neither want to understand nor live life. Past and future are equally meaningless for them, robbing them of any drive to do anything, irrespective of whether it is forward- or backward-looking.

However, if we are to understand depression correctly, it is important to realise that we are not speaking here of complete emptiness. Just as their hippocampi become stunted but do not wither away completely, people with depression are on the way towards complete lethargy and emptiness, but have not reached that destination — and that is the cause of their suffering. If they had reached it, their condition could be compared with the meditative immersion of a Zen philosopher, and they would not require treatment. But people with depression are conscious of the fact that they can still achieve effects; the problem is that those effects are judged to be negative.

They can get up in the morning, get dressed, have breakfast, go to work, eat lunch, talk to people, return home in the evening, go to bed — but all is struggle since nothing they eat tastes good, going to work seems pointless, and sleep is not restorative. And they see people around them achieving things, possibly achieving far more than they themselves — for example, by taking care of them and other depression patients. Every day, their brains experience the fact that they could be achieving effects, but they do not have the necessary drive.

The experience of helplessness in the face of a threatening event that is anchored in their memory leads to an almost automatic expectation of disaster. They are no longer driven forwards by positive expectations and objectives, which is mainly the work of the dopaminergic regions of the basal ganglia, and they enter a state of profound resignation.

This fact that people with depression are always processing towards emptiness but never achieve it should not lead us to reject all kinds of treatment for them. Current therapies worth examining include the approach taken by some psychotherapists, who give their depression patients constant positive encouragement. This might instil the feeling that their depression has helped them achieve something after all — namely, unconditional encouragement from their therapist. But why, then, should they cast off their depressive behaviour?

The commonly used pharmaceutical drugs also have little effect, and some of them can be downright dangerous. In spring 2005, the American and European drug authorities, the FDA and the EMA, reported suicides, aggression, and impulsive risky behaviour in connection with antidepressant medication. One year later, British scientists published an overview study of paroxetine, a particularly popular antidepressant. The study showed a doubling in the frequency of hostile acts under the influence of this drug; in children with obsessive-compulsive disorder, such acts increased by a massive 17 times.

Figures like this are not surprising, in view of the incomplete process towards emptiness in depression patients just described. Antidepressants work by increasing patients' drive, but they do not stop them from continuing to view their world and the people around them as meaningless. Such drugs do nothing to improve the ability of the hippocampus to package the elements of the world into a meaningful whole. Quite the opposite, in fact! Like sleeping tablets that disrupt deep sleep and dreaming, antidepressants upset the memory-forming process even more.

The most effective treatment for severe depression is electroconvulsive therapy. A strong electric current provokes an epileptic seizure, which 'dispels' entrenched negative thoughts (those stored extremely well by the hippocampus), causing amnesia, at least temporarily. It cannot be repeated too often, since the artificially induced seizures cause more than just negative thoughts to disappear: memory impairment is one of the most common side effects of electroconvulsive therapy. However, it does prevent depressive behaviour triggered by negative memories in the brain for a certain amount of time. This is similar to cognitive therapy in which confronting patients with the absurdity of their negative expectations enables entrenched memories of trauma, separation, etc. in the brain to be reshaped by replacing them with new memories (and new connections in the brain).

Psychotherapeutic, electrical, and pharmaceutical intervention all have their limits in the treatment of depression. For this reason, rather than stigmatising depression out of hand as a disease, it makes more sense to see it as an evolutionary adaptation — a rather error-prone one, but still an adaptation. This is the view advocated by the American evolutionary psychologists Paul Andrews and Anderson Thomson.[3] They believe that depression would have disappeared long ago if it were really as bad for us as we assume. But it has not disappeared, and so Andrews and Thomson believe that, at least in its non-suicidal forms, depression must have an important role to play in the survival of our species. They believe its purpose was to remove people from social cooperation and conflict situations in order to save energy.

Support for this theory can be found in an international study that showed that people with depression are more painstaking in many decision-making processes and are also more successful in the end. For example, in a role-play situation where they took the part of human-resources manager, they were able to choose the best candidate for a job, while their non-depressed peers often picked the wrong candidates. The depressed test subjects looked behind the mask, questioned candidates more closely, and tested their credibility more rigorously. And they were naturally able to see possible negative outcomes more clearly.[4]

So it may be possible that depression does not necessarily incline towards disaster, as is often claimed, and as is illustrated by the tragic fates of people like David Foster Wallace. Rather, it appears that depression was to a certain extent desirable in the history of humankind — as a condition of emptiness allowing people to stop and pause for a while.

Anything but boredom: fidgets and psychopaths

The continued existence of attention-deficit hyperactivity disorder (ADHD) and psychopathy is probably similarly thanks to some advantage conferred in the survival of our species meaning they were not weeded out by natural selection. There are indications of this in the biographies of many prominent figures in world history. Albert Einstein, Thomas Edison, Leonardo da Vinci, Hermann Hesse, Wolfgang Amadeus Mozart, and John Lennon exhibited clear signs of attention deficit, but it was probably this lust for new and powerful stimuli that made them such driving forces for innovation in their field.

We could draw up a similar list of psychopaths, since that group is by no means restricted to courtroom docks and jails, as is widely assumed. Recent studies have shown that psychopaths are represented three to four times more often in positions of power than among the general public; the proportion they make up of business executives is also around six times higher than that of the general population who are psychopathic. It is in such positions that psychopaths can best live out their fearlessness and lust for sensation, and, with their personality characteristics — such as lack of empathy, recklessness, brutality, and adventurousness — they naturally possess all the aptitudes needed to climb the career ladder in competitive societies.

However, even if both ADHD and psychopathy may have provided positive impulses in human evolution, it still seems strange, at first glance, to deal with the two conditions in the same chapter. The former conjures up images of Fidgety Philip from the children's book *Struwwelpeter*, while the latter makes us think of violent criminals and mass murderers whose brutal and calculating behaviour frightens and shocks us. Yet the two conditions share many common characteristics.

Many teenagers with ADHD are conspicuous for their risky and sometimes illegal behaviour, such as driving without a licence, shoplifting, drinking, or taking drugs — all things also typically found in a psychopath's career history. A study involving the inmates of a Scottish prison found that 23 per cent showed clear indications of a childhood characterised by ADHD, and among those who committed sex crimes or crimes involving robbery — usually the domain of criminal psychopaths — that figure rose to 31 and 35 per cent respectively.

Einstein is once again a good example of the blurred borders between ADHD and psychopathy. As a schoolboy, he was erratic and impulsive; later in life, especially in the context of his first marriage, he exhibited significant psychopathic tendencies. For example, he initially gave his wife Mileva the nickname LSD (short for *Liebes Süßes Doxerl* — 'dear little dolly') to underscore his dependence on her, but later dictated a contract to her in which he laid down paragraph after paragraph of rules for how she was to behave as his wife. She was to bring to his room 'three meals a day, as is proper', but she was to expect no tenderness from him and was never to criticise him. Furthermore, she was to leave his study and bedroom immediately and without protest whenever he wished.[5] Friends of the family were appalled at Einstein's cold-hearted treatment of his wife once he was no longer interested in her.

The fact that the boundaries between ADHD and

psychopathy are blurred is ultimately supported by the brain's physiology. Both conditions include underactivity in and lack of fine adjustment (connectivity) between the cingulate gyrus, the amygdala, the insula, and the ventral striatum on the one hand, and the prefrontal cortex, which is responsible for self-control, on the other. The complex of functions carried out by these regions can be described as the triumvirate of control over attention, drive, and emotional self-regulation, which already points to the consequences of a lack of fine adjustment in that respect:

- low susceptibility to fear
- no avoidance of negative consequences for self or others
- sensitivity to monotony and boredom
- emotional dysregulation, i.e. marked mood swings and irritability
- disturbance of social perception and empathy, little compassion, and ignoring or misinterpreting the reactions of others
- delay aversion, i.e. a desire for immediate positive reinforcement and an unwillingness to wait for rewards.

Results of this intolerance towards delayed gratification include a propensity to seek out sensation, i.e. to strive for intense stimulation for immediate gratification. Harmless iterations of this behaviour are constant channel-hopping while watching TV or constant googling on the internet, bungee jumping, rock climbing, competing in triathlon, and expensive shopping sprees. Less harmless expressions of this behaviour can include gambling on the stock market, involvement in fraud scams, indebtedness, sadism, animal cruelty, manslaughter, and murder. The important thing is that the behaviour must deliver a kick.

It is the search for new source of stimulation that ultimately categorises ADHD and psychopathy as disorders of emptiness, since what such patients fear most is monotony and boredom. Their brains cannot stand it when nothing is happening. Although, as described in Chapter 1, this is also true of many other people, it is compounded in ADHD patients and psychopaths by the fact that they find far more things, indeed virtually everything in our daily lives, boring.

The threshold at which positive arousal is reached is relatively high in such patients, and it is raised even higher through experience. While it might be enough at the beginning to talk back to the teacher during class, later the same effect can only be achieved by kicking her in the shin; while it might be enough at first to stand at the edge of a cliff with no safety barrier, later the same effect can only be attained by BASE jumping off the cliff; and while it might be enough at first to hold a lighter to a cat's tail, later the desired kick can only be had by throwing an entire hutch of rabbits onto the bonfire. This escalation means in turn that day-to-day stimuli are perceived as increasingly boring.

Thus, many ADHD patients and psychopaths have a particular problem: in their battle against emptiness they automatically increase the extent of that emptiness. It's a vicious circle, which should be taken into consideration in psychopaths especially, since they are known to lack any compassion. This does not excuse their behaviour, but, compared to their hopeless railing against emptiness, the mountainside escapades of Sisyphus look almost like a stroll in the park.

Yet this doesn't mean that psychopaths are lost forever. In our experiments, we succeeded in training prison inmates with severe psychopathic tendencies to deliberately activate the under-functioning parts of their brains and thereby alter their patterns of behaviour. Also, less severe cases succeeded in coming to terms with their inner emptiness to such an extent that they were able to avoid slipping into a spiral of escalating stimuli. The US criminal psychologist Adrian Raine also comes to this conclusion in a study that was remarkable not only for the results it yielded, but also for its design.[6]

Raine was interested in finding psychopaths who were not in prison, and who had therefore demonstrated an ability to navigate day-to-day life more or less successfully. He could have turned to top executives in companies, organisations, or political parties, but they were unlikely to have either the time or the inclination to participate in a psychological study. So Raine hit upon the idea that temping agencies might be a good source of such people. Such agencies may not be a place to earn high salaries, but otherwise they offer everything a psychopath could want, including short-term goals and success, ever-changing challenges, and short-lived social contacts with little commitment. And, indeed, when Raine examined temp-agency workers, he found eight times more psychopaths among them than the average population.

Raine's examinations revealed an important difference between socially successful psychopaths and those who had fallen foul of the law. The law-abiding psychopaths had a much lower heart rate and sweat-gland activity at rest. Although their involuntary functions were reactivated all the more rapidly and powerfully as soon as any intense stimuli appeared — watching an action movie, for example — societally established psychopaths are clearly better able to deal with inner emptiness, since they would otherwise develop much more pronounced stress reactions.

By the same token, this might mean that cultivating a level of comfort with inner emptiness provides some protection against the development of psychopathic criminality.

Borderline: detached, but not free

All was well in the life of Dan Gallagher, a lawyer in New York. His career was going well, and his family life also appeared perfect. But then he started a passionate affair with a publishing editor called Alex. For him it was clear that it was just a fling, and nothing more would come of it. He didn't want it to interfere with his good life. But Alex had different ideas.

Even direct rejections were not enough to stop Alex's desire for Dan from becoming increasingly obsessive, or to prevent her from putting pressure on him. She claimed to be carrying his baby, tried to kill herself, and eventually kidnapped his daughter — after first killing the daughter's pet rabbit. The story ends with a big, bloody showdown, in which Dan's wife shoots the obsessed stalker.

Well, that's how the American version of the film *Fatal Attraction* ends. The Japanese version of Adrian Lyne's movie ends as the original screenplay intended — with Alex committing suicide. The producers decided this would be too much for Western audiences to accept, but it was kept in for viewers in Japan, where there is a tradition of honourable suicide.

The movie, starring Glenn Close and Michael Douglas, was released in 1987 but even today it is considered a cinematic masterpiece in the portrayal of borderline personality disorder.

Hundreds of millions of people are thought to suffer from borderline personality disorder. One of the main symptoms is an inability in those affected to regulate their emotions. This means that the threshold above which they become emotionally agitated is particularly low, while the level of agitation is higher and emotions are experienced as particularly intense and cannot be modulated. People with borderline personality disorder can usually see things only as black or white, good or evil, and so — like Alex — they are unable to accept someone having sex with them without wanting to enter into a relationship and break up their family. In the same way, they cannot bear abandonment of any kind, because they interpret this not as a temporary situation, but as a life sentence to loneliness. And this seems to indicate that emptiness plays a major and frightening part in their lives.

This assumption is supported by the fact that people

with borderline personality disorder themselves often report an unbearable feeling of inner emptiness. More than half of patients treated say the constant emotional rollercoaster they are on robs them of any certainty about 'who they really are'. Which brings us to a central feature of their disorder: dissociation. This is a sense of detachment from oneself and one's own body. People with borderline personality disorder are somehow 'not themselves'; they experience their own behaviour as not being under their control. Rather, they observe themselves from outside, like an object, and the cause for this is usually a childhood trauma.

There is no other personality disorder in which we find so many indications of physical abuse in the — mostly female — patients. According to investigations in the USA, more than half of all borderline patients treated in hospital had experienced incest. The relationship between the perpetrator and the victim is important here. Perpetrators of incest are not just anybody; they are an important — often the most important — attachment and trust figure for their victims. This means victims can no longer view their fear as a result of violence from outside, and they have no possibility of seeking protection, because the person who should be protecting them is the aggressor.

Since they can find no object either for their fear or for their search for protection, victims are forced to find another coping mechanism, and that consists of 'stepping out of themselves' as a way of removing themselves from immediate involvement and having to deal with their lives. Such dissociation becomes a lifebelt for them in the stormy sea of traumatic experience. Many borderline patients also tend to self-harm, by cutting themselves for instance, because this dissociation from their own bodies means they feel physical pain less keenly. But by reducing themselves to the level of an object by self-harming, people with borderline personality disorder can blur the borders between themselves and other objects; they often overstep borders in their dealings with other people.

At the Max Planck Institute for Human Development in Berlin, scientists wanted to find out how the brain activity of borderline patients changes when they are shown photographs of people displaying positive or negative emotional reactions.[7] The findings showed that such patients developed much stronger activation of the insula when compared to a healthy control group.

That part of the cerebral cortex is very old in evolutionary terms, and its function is to inform us of our current emotional state by reporting to other areas of the brain the bodily changes caused by our emotions. And because it also allows us to recognise similar physical changes in others, it makes us capable of empathy. In psychopaths, the insula is usually barely active at all, but in people with borderline personality disorder, by contrast, it works overtime. This might lead us to view these two disorders as each other's opposite, along the lines of: what one has too little of, the other has too much of. Only, people with borderline personality disorder are not really empathetic at all.

As the team of Berlin researchers found, such patients

do show a strong reaction to seeing emotionally charged photos, but this reaction cannot be called empathy, in the sense of feeling what another person feels. Borderline patients' brains are so occupied with their own feelings and their — mostly unsuccessful — attempts to control them, that they have no real ability to take on the view of others. It is no coincidence that borderline patients have also been found to have diminished amygdalae and hippocampi, both of which play a central role in controlling negative emotions.

For this reason, the lead researcher in the Berlin study, Isabel Dziobek, prefers to speak of 'emotional contagion' rather than empathy. This contagion is so strong that 'the borderline patients are no longer in a position to engage with the person they are seeing'. They did feel something when they saw other people in emotionally charged situations, but it was not the same emotion felt by the person in the picture. This finding matches observations from clinical psychologists that people with borderline personality disorder often ascribe their own feelings to other people, and, for example, project their own anger onto their partner ('You seem so angry today').

Treatment of borderline personality disorder is difficult. Its success depends to a great extent on whether the early environment of the person concerned was particularly brutal, or whether it involved a violation such as father-daughter incest. Many patients abandon treatment because they cannot deal with their therapist and — true to their propensity to see everything as black or white — either adore or despise him or her.

However, that doesn't mean there's no real chance of success in treating borderline personality disorder. Experts estimate there is in fact a 60 to 70 per cent chance of improvement. This is probably not down to treatment exclusively, but also to the fact that borderline personality disorder does not involve a complete loss of the patient's grip on reality, and so leaves room for a certain amount of therapeutic success.

Yet when it comes to people with schizophrenia, that border has long been crossed.

Schizophrenia: a flood of stimuli into nothing

Did you know that the two medical terms 'schizophrenia' and 'dementia' have shared historical roots? When the Swiss psychiatrist Eugen Bleuler first introduced the term 'schizophrenia' to the world, at a meeting of the German Association for Psychiatry on 24 April 1908, he already suspected that it would have difficulty establishing itself among his fellow psychiatrists.

The term comes from Greek, of course, and means literally 'a splitting of the mind'. How can a split mind be healed? Compulsions, anxieties, tics, and depression — all those can be treated. But a split mind? How should psychiatrists deal with an illness they cannot cure?

So Bleuler also fell back on a term that was common among psychiatrists at the time: dementia praecox. When he published his seminal report in a journal of psychiatry, he gave it the title 'Dementia Praecox, or the Group of Schizophrenias' [*Die Prognose der Dementia praecox (Schizophreniegruppe)*'].[8] The problem was, the German psychiatrist Emil Kraepelin had already claimed the term 'dementia praecox' to describe a state of depressive indifference and stupor, which is much closer to our modern usage of the word dementia.

Conflict was in the offing. But that was no problem for Bleuler. He once said, 'The universal curve of the entire world would be different if I had not lived.' Anyone with that kind of self-confidence — almost bordering on the megalomania of his schizophrenic patients — has no fear of conflict. Bleuler stood by 'his' dementia praecox. This presented his fellow psychiatrists with the problem of deciding how they should refer to the disease. They excluded 'dementia praecox' out of respect for Kraepelin.

However, there was no need to avoid the term 'dementia praecox'. We often imagine people with schizophrenia to be raving lunatics tearing up the asylum, but there are other forms of the disease, such as catatonic schizophrenia and hebephrenia, which are characterised by depressive indifference and stupor. That is, reminiscent of the dementia patients we see in care facilities today. Moreover, dementia and schizophrenia are both connected with a certain area of the brain that we encountered when we considered depression and borderline personality disorder: the hippocampus.

A research team led by Jessica Turner of Georgia University in Atlanta analysed MRI brain scans from more than 2,000 schizophrenia patients and just over 2,500 healthy controls, gathered at 15 centres.[9] This analysis revealed a significant reduction in the size of the mentally ill patients' amygdalae and nuclei accumbens, which are responsible principally for the negative emotional marking of information, but also for the production of feelings of happiness. Yet the reduction in hippocampus size was the most striking: it was an average of 4 per cent smaller than in healthy patients. As we already know, this is the area of the brain where individual memory items are combined to create a meaningful whole, so it is the location of an important part of our associative abilities. A reduction in its size may be interpreted as a step towards dissociation, and thus also a step towards emptiness; the exterior, social world loses its significance and becomes fragmented — which is evidenced in particular in the behaviour of schizophrenic patients.

Even long before the onset of their illness, people with schizophrenia often display a conspicuous inability to differentiate between significant and insignificant information. Normally, the brain works like a fisherman, who only takes from his net the catch he can use: the fish go to market; the old tyres and tin cans go back in the water. In a similar way, only significant signals are processed by the brain and the rest are ignored. That saves effort and energy. However, this mechanism no longer works in the brains of people with schizophrenia, which gather the trash from the net along with the fish — meaning they do not filter the signals they receive, but process everything without any kind of triage.

This explains why the thalamus, which normally carries out that filtering process, is also less active in schizophrenic patients. This leads to a barrage of stimuli, which can open up whole new worlds and fresh perspectives, triggering remarkable creativity, as in the case of the poet Friedrich Hölderlin or the mathematician John Nash.

But such an overabundance of stimuli can also lead to precisely the opposite: nothing. If everything is equally significant, then, ultimately, *nothing* is significant. It is no coincidence that Nietzsche sought 'the destruction of all morals' as the basis of nihilist philosophy. Thus, their inability to confer meaning on things inevitably brings people with schizophrenia close to emptiness: meaningful relations can no longer be made.

Part of this is the common feeling among people with schizophrenia that they are being remote-controlled. If everything is meaningless, there is no longer any differentiation between the internal and the external, between the self and the outside world. Some 84 per cent of psychotic patients perceive thoughts that they believe originate from outside, although they are of course produced by themselves. The typical case is hearing voices, which are rarely commanding in nature, but whose constant presence alone influences the patients' behaviour. People with schizophrenia often feel the voices are abusive or even threatening towards them. Some psychotics shoot at harmless birds because they feel the creatures are pursuing them. Others fly into a rage when someone gives them a friendly smile, because what they see is a scornful sneer. And just as even the most composed of people will eventually react angrily to constant abuse and persecution, people with schizophrenia will do the same. Only, for people with schizophrenia, even weak negative stimuli are enough to provoke that reaction.

The feeling of being controlled by an exterior force can have far-reaching consequences. From the sense of being manipulated like a puppet on strings, the brains of people with schizophrenia learn that their own actions no longer achieve effects. When no effects can be achieved, it no longer makes sense to *want* to do anything. This is why schizophrenia patients eventually descend into a state of indifference and stupor. During that process, many have the feeling that nothingness is descending inexorably over their lives like a menacing shadow. Such a feeling naturally results in a state of fear and often a wish for it all finally to come to an end. Up to 10 per cent of schizophrenia patients eventually kill themselves. However, most do so as they are descending into emptiness, and not when they have already arrived. And that brings us to a crucial point.

The fact that the tendency of schizophrenia patients to want to kill themselves decreases as their condition progresses could be seen as being due to a lack of strength and the necessary cognitive abilities in the later stages of their illness — after a long period of suffering. But there is another possible explanation: while they are still in the process of descending towards emptiness, people with schizophrenia still have the capacity to realise that they are losing the life they have known. Naturally, such a realisation causes feelings of desperation. But later, when they have arrived at a state of emptiness, their will is extinguished and along with it the urge to try to change anything. Suicide now seems just as meaningless as everything else, and so there is little danger of it being realised. Against this background, schizophrenia patients could even be considered to be 'liberated': they no longer have to initiate anything, not even their own death.

We would not go so far as to describe the final state of emptiness in schizophrenia patients as perfect happiness. But we would also not describe it as unimaginable suffering, either, simply because it leaves patients apathetic and lethargic. Later, when we look at locked-in patients, we will see even more convincing evidence that a situation that we, as individuals capable of moving and wanting, simply cannot imagine, should not simply be dismissed as a living hell from which the patient should be freed as quickly and 'humanely' as possible. Emptiness is a state our brains can certainly cope with, even if the journey there is often considered tragic. The terrible thing about mental illnesses such as depression, borderline personality disorder, and schizophrenia is that they send sufferers on a long, slow slide into emptiness. They have little or no control over that process, yet realise nonetheless that they are losing their former healthy lives forever.

But what happens at the bottom of that slide, when the journey is at an end? The fates of dementia and Alzheimer's patients show that it need not necessarily be a living hell.

Dementia: I'd rather sit in the corner drooling

Enough of work! When Theo retired, he decided he was not going to do anything anymore. And he really meant absolutely nothing! Travelling to the German Baltic Sea island of Fehmarn to spend summer at the family holiday home there, but nothing else! Former workmates tried to persuade him to join their men's choir since he was a good singer, and many a charity would have been glad of the former banker's voluntary services as a bookkeeper. But Theo was having none of it. No amount of persuasion was going to change his mind, and, when he was warned that inactivity was a one-way ticket to dementia, it just made him even more obstinate. 'I'd rather sit drooling in the corner than break my back slaving away for someone anymore.' He had worked his way up the ladder in post-war West Germany, starting as a penniless refugee from the Soviet-occupied former Prussian territories and ending up on the board of Deutsche Bank, and had seen and understood the way work and exploitation often go hand in hand. And so, when he entered retirement, he was adamant about one thing: I'm not working for anyone anymore.

When Theo was in his mid-70s, the family sold their island holiday home. From then on, he alternated only between the kitchen and his bedroom, no longer left the house, and gained more and more weight. He barely watched television ('They only ever show crap, anyway'), he had never been much of a reader ('It just ruins your eyes'), and he consistently ignored his hearing aid ('I can endure the world better without it'). But the threatened dementia did not initially materialise. It was as if his brain wanted to show everyone that it needed nothing more than just itself. It was not until he turned 80 that Theo became increasingly forgetful, his natural urge to move disappeared almost completely, and he became incontinent and occasionally aggressive. In short: he was care-dependent. His wife looked after him at first, but she eventually had trouble coping with the 120-kilo man, and he was placed in a home.

The first few months weren't easy, because Theo still held out the hope of eventually returning home. He sometimes lost his temper, but his anger ebbed away with time. He blocked out the care home in its function as a care home — for him, it increasingly became a hotel. On the island of Fehmarn, apparently, where the family had had their holiday home.

Three weeks before he died, Theo received a visit from his youngest son. He was happy to see his offspring, and his son was happy that his father still recognised him. Theo asked, 'Where are you putting up, where are you staying?'

'With Mama, at home. It's not so far from here.'

'You can't be. We sold that place.'

'Erm … Actually, Mama still lives there.'

'Nah. Can't be. We sold it.'

The son was bemused, but for the father, the subject was finished.

After a while, he asked, 'Shall we go to the beach later and then go for something to eat? A nice piece of fish, or something?'

The son realised that his father thought he was on Fehmarn. So he answered, 'Okay, let's do that.'

Theo was satisfied with that. He turned away and watched the pigeons through the window. After a time, he turned back and expressed delight that his son had come to visit him: 'Where are you staying?'

'At the high-rise hotel on the beach front. The one with the three towers.'

'Ah, yes. How much is it?'

'Eighty euros a night.'

'What? They must be crazy.' After a little more chit-chat, he turned away again to watch the pigeons. After a few minutes of silence, he turned back again and once again was delighted to see his son had come to visit: 'Where are you staying?'

'At the high-rise hotel on the beach front. The one with the three towers.'

'Ah, yes. How much is it?'

'Eighty euros a night.'

'Oh. That's not bad. I'd've thought it'd be more expensive.'

Theo passed away shortly before his 88th birthday. He died in his bed, and the carers at the home were sad to lose one of their most amicable and witty residents. In the final few months of his life, he had found peace. And that had nothing to do with the infamous joke about people with dementia being lucky because every day is a new beginning with no memories to burden them. No, it was because things lose their meaning for a brain with dementia. Eighty euros for a hotel room can be cheap or expensive — still worth mentioning, but not important in itself and not connected to very many associations or very much meaning. The son can have just arrived or have been sitting there for hours — it doesn't matter either. No more responsibility for others; no more heavy, burdensome meaning; just gentle emptiness — and the source of that development lies once again in the hippocampus, one of the oldest parts of our cerebral cortex.

Dutch researchers used magnetic-resonance imaging to examine the brains of healthy people and those with dementia, and found significant differences in the size of their hippocampi. But they also made an even more significant discovery: those among the healthy control group who already had a smaller hippocampus when they were first scanned were six times more likely to develop dementia within a year and a half. And if their hippocampus continued to shrink in those 18 months, their risk of dementia leapt to 61 times the normal rate. For the lead scientist of the study, Wouter Henneman of the VU University Medical Centre in Amsterdam, this is clear evidence that the extensive loss of brain matter often seen in images from Alzheimer's patients is a sign of the advanced stages of degenerative brain disease, 'but it begins in the hippocampus'.[10]

As already described, the hippocampus is the area where individual items of memory are bundled together to form a meaningful, comprehensible whole. In the case of degenerative shrinkage, this means that our world of thoughts and perceptions becomes devoid of any context. This is evidenced by the further progress of the disease and can also be seen in other areas of the brain.

For example, it leads to a dramatic loss of substance in the synapses of the cerebral cortex, presumably due to the degeneration of the associative hippocampus. The result of that loss is that increasing numbers of individual nerve cells become isolated from each other. By the time around

10 per cent of a brain's synapses have perished, its owner will usually be displaying clear signs of dementia.

Doctors use medicines called cholinesterase inhibitors to try to slow this synaptic decay. The drugs work by slowing the breakdown of acetylcholine, an important brain messenger substance for associative connections. Experiments have also shown that these drugs increase the number of mitochondria, which means nerve cells have more energy. Their clinical success has been moderate, however. Germany's Institute for Quality and Efficiency in Health Care estimates that the use of cholinesterase inhibitors leads to short-term cognitive improvement in just 16 per cent of dementia patients, and improvements in coping with the demands of everyday life are estimated for only 8 per cent of patients.

It is also questionable whether non-pharmaceutical methods such as so-called brain-jogging exercises can slow down the progress of dementia. A committee of experts under the leadership of the Stanford Centre on Longevity and the Max Planck Institute for Human Development analysed relevant studies and came to the conclusion that brain teasers and software-based training programs only improve the skills that they directly train: 'However, only very few programs had a positive effect on overall mental abilities and performance in everyday situations.'[11] Training the memory by memorising lists of words leads to an improved ability to memorise lists of words — nothing more.

The loss of synapses due to dementia appears to be impossible to stop. This may be due to our ever-increasing life expectancy making too many demands on our brains' regenerative abilities. Apes, which die at the age of 40 or 50, do not suffer from dementia. Yet this should be no cause for panic. When we think of dementia and Alzheimer's, we often make the mistake of focusing only on the losses those diseases inflict. But we can also gain something from them: emptiness.

In order to understand this seemingly illogical gain, let's take a look at how the world inside our heads is composed, as long as the brain stays fit. It is made up of associations — that is, connections between A, B, and C, etc. For example, we recognise our mother because she has the right voice and appearance, and we board a bus in the confidence that it will take us to work because we recognise the number on the front. These are all associations that we piece together out of the individual perceptions we gain of the world around us, which we are thus able to navigate through and orient ourselves in. That world becomes increasingly lost, however, as dementia progresses.

While individual regions of the brain remain perfectly functional — as can be seen in EEG readings — they become increasingly unable to communicate with each other. As a result, the perceptions we have of things in the world around us become increasingly 'isolated'; they lose the meaning that they formerly gained by virtue of their relationship to one another. Initially, a bus can still be recognised as a bus, but we can no longer identify where it is going. Eventually, even the association between the thing and our concept of it breaks down, and the bus is no longer recognised as such. This is the stage at which dementia patients take their leave of the world of concepts.

It is clear that such a development is at first experienced as an agonising loss, since it tears us away from the world we are familiar with. The safe orientation points of our associative consciousness are gone, and that is frightening. Not to mention the fact that we can tell that other people are *not* subject to such a development, causing an additional feeling of inferiority. Other people can remember last night's TV show or their last holiday as a matter of course, while we can't even remember what we had for lunch an hour ago. Our circle of friends and family grows ever smaller, because we simply no longer remember who those people are. That's tough.

But, with time, those negative feelings grow weaker, because, as the meaning of things disappears, so does the ability to compare them. Dementia patients progressively lose the feeling that there is something wrong with them in comparison to before or in comparison to other people, because they are simply no longer able to perceive the difference.

Eventually, then, dementia patients no longer suffer under the circumstance that they are drifting away into emptiness. First, because they have become detached from an existence filled with meanings and have grown used to that condition rather than being alarmed by it. Second, because emptiness, like many other things, is ceasing to exist for them as a concept. This is why most dementia patients eventually see it as a nuisance when someone tries to drag them out of their dozing state.

Anyone who tried to encourage Theo to join in with bingo, solve sudoku puzzles, or play any other kind of games received a gruff snub. He still always wanted to have his daily paper within reach, but really only because it was part of his familiar environment — in his final years, he never read a single word out of it. Whenever someone suggested he might want to have a leaf through it, he would either tell a little lie ('I've already read it') or snap, 'Why don't you read it, at least then you'll shut up.'

We should not judge someone by the standards of our lives when that person has left such a life behind them. At Tübingen, we were forced to admit that this is often a mistake when, during one of our studies, we tried to make contact with dementia patients who were barely communicative in day-to-day life. We chose to use the classic conditioning approach first developed at the end of the 19th century by the Russian scientist Ivan Pavlov. We played various sounds to our test subjects and associated them with an immediate positive or negative stimulus, such as the sight of a delicious snack or a piercing tone (prior consent was received from the subjects' families or legal guardians where necessary). At the same time, the subjects' brainwaves were recorded using EEG, and the blood flow to their brains was measured with MRI. The resulting information was analysed by a computer. We continued this procedure until the computer was able, using only the brain activity data, to recognise reliably when the subjects' emotions were negative or positive. We then exposed our test subjects to pleasant or unpleasant images, for example a verdant landscape or the scene of an accident or an act of war. In many cases, the subjects were unable to describe the scenes, especially if they had never experienced anything comparable in their own past lives. Nonetheless, the computer was able to tell from the subjects' brainwave patterns whether they had a positive or negative reaction to the images. Thus, we had made our dementia patients talk, albeit at a very simple emotional level.

We were soon imagining that dementia patients would be able to use this method to communicate — to ask whether they found certain people, or life in the care home, unbearable. That would not only be a relief for those residents. It could be a source of embarrassment for those around the patient who had insisted she must be happy because she never complains. But reality was soon to bite. We found that our technique did not work for people with advanced dementia. Either the computer was unable to recognise any logic in their brainwave patterns, or those test subjects did not display an unambiguous reaction to the images.

We abandoned the project because it was not helping us move any closer to our goal of communicating with people with advanced dementia. However, in retrospect, I would say that study did provide us with an important insight.

If a computer can no longer recognise any emotional logic, that may be seen as a clear indication that a patient has left behind the world of meanings as we know it. Associating things with positive or negative feelings ultimately goes the same way as all other associations. As healthy people, we see that as an emotional impoverishment and therefore a heavy loss. This explains our tendency to want to provide dementia patients with positive experiences at all costs, despite the fact that they often might not want them, since for them everything more or less descends into empty meaninglessness, leaving them *unable* to want anything at all.

One time, Theo's family decided to take him out for an ice cream. It turned out to be a disaster. Theo wanted to stay in the care home: 'There's ice cream here, too.' When his family argued that it is nice to eat ice cream out in the open, his reply was simply, 'Not for me.' And he was right.

Patients with advanced-stage dementia no longer harbour the same values as we do; they do not find the same things pleasant or unpleasant. When things lose their meaning in the world of advanced-dementia patients, those people also take their leave of the value system we operate in. This does not mean we should stop caring for dementia patients. Sometimes they re-emerge totally unexpectedly from their emptiness.

Once, one of Theo's fellow residents struck up a song that was familiar to the others in the home, and suddenly almost everyone was singing the old favourite with gusto. It was a choir of people who had let everything go and were clearly enjoying themselves. When they had sung through all the verses and the carer asked if they wanted to sing another song, the residents all suggested the same one again. And then again. Because those old people were not looking for variety, but rather continuity. Then silence descended on the room once again. It was as if it had never happened. Later, Theo was unable to remember the impromptu singing session. It was nice while it lasted, but not worth mentioning otherwise. It is in the nature of emptiness that it only allows moments, and, once they are gone, they are gone.

11

The Right Life in the Wrong Body[1]

the happiness of locked-in syndrome

Waltraut Fähnrich is 70 years old, but she doesn't look it. Her face looks young, almost girlish. Her skin is waxy; there is no sign of the wrinkles the faces of others of her age usually bear. But no one would say that Waltraut is spritely for her age. Although she doesn't have pains in her knees or back, that's because she is no longer able to do anything that might harm her joints. And her youthful complexion isn't thanks to bracing country walks or a good workout program. It's one of the typical symptoms of the illness she has: advanced-stage amyotrophic lateral sclerosis (ALS). Her brain no longer has any contact to her muscles, as its motor neurons have been irreversibly destroyed. This youthful-faced woman can no longer walk, dance, grasp, or eat. Indeed, she can't even breathe without assistance. She is completely locked inside her own body.

When she has an itch, she cannot scratch it; when a fly annoys her, she has to wait until someone else shoos it away. Waltraut's eyes are still open, but she is forced constantly to stare in one direction since her muscles no longer work. Her view only changes when someone moves her wheelchair; and no one can say whether she prefers the new vista to the old one.

How could anyone find anything about Waltraut's condition pleasant? We instinctively feel horror at the thought of being locked-in ourselves, and, when we see others in that condition, we wish for their sake that their suffering will quickly come to an end. It is this fear of being fully conscious but trapped in their own body and unable to act on their own will that prompts many people to draw up an advance healthcare directive, or living will. Surveys carried out among the family members and doctors of ALS patients appear to confirm this view: 90–95 per cent feel the patients' lives are no longer worth living.

However, it is worth taking a closer look at the locked-in state. The surveys cited were carried out at a point in time when the ALS patients were still going through the painful experience of progressively losing possession of their own bodies. It is clear that patients despair in this phase. They see others moving normally through the world while they gradually lose access to it. Eating, playing sport, having sex, or dancing — they can no longer do any of the things they used to enjoy. Not to mention the fact that they lose all privacy because they require round-the-clock support from carers. Their lives become dominated by a painful and unstoppable loss.

But what about the period after that phase is over and they are completely locked in? Would they still answer 'no' when asked if their lives are worth living? The problem is how to elicit meaningful answers from someone in that stage of their illness. They can no longer speak or write, and gestures and facial expressions are also a thing of the past for them.

The famous physicist Stephen Hawking communicates by twitching a muscle in his right cheek, and those motions are picked up by a tiny infrared sensor in his glasses. The sensor is connected to a computer that translates his cheek movements into words, and they are then spoken by a speech synthesiser. It can take ten minutes for Hawking to formulate a single sentence, making communication a slow and laborious business. But at least the physicist can still maintain contact with his environment. He has a rare form of ALS in which the paralysis is either not total or takes much longer to set in.

Within a few years, most other ALS patients reach a point where they can neither twitch their cheek nor blink their eyelids. How can we communicate with them then?

Signals out of the silence

If a brain can no longer trigger any muscle activity, then we must let it 'speak' directly to us. Those were our thoughts in Tübingen when we decided to try to establish contact with locked-in patients using EEG — that is, by measuring their brainwaves. We found that they do not produce any high-frequency 'ripples', the brainwaves typical of activity in the hippocampus. Instead, in a waking state, their brains produce waves with a frequency of 6–7 Hz, in the theta range.

Those are the slow waves we see in the 'twilight state', which we have already identified in connection with meditation and floatation tanks as playing an important part in producing a state of emptiness. This should not be surprising, since both mediation and floating share with the locked-in condition the fact that barely any sensory stimuli — especially from the proprioceptive sense of body awareness — are being processed in the brain. But there is also an important difference. We know that meditating and floating are generally experienced as positive, but both are situations that are entered into voluntarily and which can be ended any time. The same is not true of the locked-in state. Patients are cast into that condition against their will, with no chance of ending it whenever they feel like it. Being locked-in is a fate over which there is no control. So it is possible that in this state, emptiness is experienced as a senseless, gruelling kind of hell.

We felt this warranted further investigation. We had to carry on trying to establish contact with locked-in patients so they could tell us how they experience emptiness.

We trained them in neurofeedback techniques, to teach them to indicate a particular letter of the alphabet, or a 'yes' or 'no', by deliberately generating certain brainwave patterns. Although this method is extremely successful with healthy subjects and with locked-in patients who are still able to move their eyes, when we tried it with completely locked-in patients the results were demoralising. We were unable to find any indication that such people were able to perceive anything consciously, let alone communicate.

However, EEG technology relies on sensors placed on the outside of the skull. That means, at quite a distance from the brain; not to mention the fact that the cranium does not by far conduct all currents. So then we tried with electrodes implanted directly in the brain. The patients' families gave their consent to the operation since it gave them a hope of being able to communicate with their loved-ones. Yet those results were also disappointing.

Next, we abandoned the idea of letting their brains speak via electrical impulses, and decided to try changes in blood flow as a means of communication, using near-infrared spectroscopy or other imaging technology to render them visible. The 'communication advantage' of blood flow is that there are already receptors in the blood vessels of the brain that can sense it. The brain registers when blood flows though it at a higher or lower pressure. We aren't able to put it into words, like describing something we've seen or heard, but we register it unconsciously.

For our first attempt, we chose a patient who was left completely locked in by ALS: Waltraut Fähnrich. We trained her to activate certain brain regions to indicate 'yes' or 'no'. She was not required to do this consciously, but simply to think 'yes' or 'no' in answer to our questions. The next step was to ask her specific questions about her wishes and her mental state. For example, whether she enjoyed visits from her children, or whether she was in pain. And she did indeed direct blood to the areas of her brain designated as meaning 'yes' or 'no'. She was certainly thinking something, because there can be no changes of blood flow in the cerebral cortex without that happening. But we did not know *what* she was thinking in order to generate those blood-flow changes. And, to be honest, we didn't care: of all people, those condemned to complete immobility should have the right to free thought.

We then reversed the questions, asking them in the negative, so that Waltraut had to give the opposite answer to her previous responses. We did this to make sure the results were not due to pure coincidence. Our patient reacted correctly, changing 'yes' to 'no' and 'no' to 'yes'. Not always, but with a certainty of more than 70 per cent over many days — far beyond the 50 per cent chance of tossing a coin. When the question was about something that was important to her, she even answered consistently 100 per cent of the time.

Pleasant moments get better, terrible moments lose their horror

Since then, we have successfully used neurofeedback training based on visualisation technology to help other patients communicate. This usually begins with exposing patients to questions or statements whose answers or veracity are obvious to anyone: 'Do elephants have trunks?', 'People have legs', or 'Is up the opposite of down?' We also deliberately use false or absurd questions and statements: 'Columbus discovered Asia', 'Paris is the capital of Germany', or 'Elephants have feathers'. This enables us to determine how the patient's yes/no thoughts are represented. The patient and the computer create a typical pattern for 'yes' and 'no' thoughts for that patient.

Later, we ask concrete questions whose answers we don't know. Locked-in patients usually take 15 to 25 seconds to think their answer. The result is then vocalised by a computer voice: 'Your answer was recognised as "yes".' Locked-in patients can now be questioned by their relatives using this BCI (brain–computer interface) technology. Waltraut Fähnrich has been using it to communicate with her husband for quite some time.

So, it is possible to communicate with completely locked-in patients. And thus we were also able to ask them about their subjective quality of life: 'Are you pleased to see your children when they visit?' Or, more generally: 'Are you happy, on the whole?' And eventually, the question of all questions: 'Do you want to die?' Or, formulated the other way around: 'Do you want to carry on living?' We have asked our patients these questions many times and made it clear to them that if they give a stable, repeated 'no' in answer ('I do not want to carry on living'), they will be given help and we will act on a living will to that effect. Yet it appears that those who are locked-in and conscious of the fact that there is no hope of that situation changing are precisely the ones who are extremely attached to life.

Some critics pointed out that our locked-in patients may have only been pretending to love life so much in order to stay on our good side — after all, our machines and our method were their only remaining chance of communicating with their fellow human beings. So, we devised an experiment to measure their quality of life.

We used magnetic-resonance imaging because it can detect blood flow changes in the deeper, emotional areas of the brain. Patients were placed in an MRI scanner and confronted with images (for those who still had use of their eyes) or sequences of sounds (for those who could only hear) that aimed to provoke positive or negative emotions in them. Members of a control group of healthy volunteers were subjected to the same procedure.

The images and sounds were taken from the International Affective Picture System (IAPS), developed at the University of Florida by Peter Lang and used by psychologists all over the world to elicit and record emotional reactions. The picture set includes images that would affect anyone in the world more or less intensely, irrespective of their cultural background. It contains, for example, an image of a pile of human corpses and pictures of maimed children to provoke reactions of fear, horror, and disgust. It also includes pictures of naked women (for male subjects) and laughing infants (for female subjects) to provoke feelings of lust or joy. For subjects without visual perception, there is a set of corresponding sounds, including the sound of the sea and of children's laughter, as well as human cries of pain and the sound of war-time air raids.

The MRI scans showed that the severely locked-in patients had a more pronounced reaction to positive stimuli and a less pronounced reaction to negative stimuli than the healthy control group. And those response patterns were stronger, the more advanced a patient's illness was. Thus, there was no sign that these severely disabled people with highly limited means of perception were able to react only scarcely to their environment. Contrariwise, there was also no indication that their lives are so extremely uneventful that they were more intensely shocked by negative stimuli than their healthy peers. Rather, that which makes us happy, made these people even happier, and that which makes us unhappy affected them much less. Ultimately, the conclusion must be that their quality of life is higher than ours. Of depression or resignation not a sign!

It was striking that the locked-in patients displayed stronger activity in the supramarginal gyrus, a part of the cerebral cortex found in both hemispheres where the parietal, temporal, and occipital lobes meet. When this area is active, it sends blocking signals to the amygdala and other parts of the defence system. We met this system in Chapter 4 in the context of scaling it back as a basis for a positive experience of emptiness. In normal life, we often fail to reach that state, and the result is that the things around us take on too much — mostly negative — significance, which prevents us from achieving the meaninglessness necessary for emptiness. This is why we have to fall back on sex, music, meditation, floatation tanks, and other 'emptying techniques' to be able to have that experience, at least temporarily. In the case of locked-in patients, however, all that is behind them. Emptiness comes to them without their having to seek it out. But, more importantly, it makes them *happy*.

Hopelessness at last

An inside view of the brain is not the only thing that can help us take a much more positive view of the fate of locked-in patients than a superficial assessment might give us: lessons from behavioural science and research into psychological resilience (the ability to deal with crisis situations), as well as ideas philosophy, can help us, too.

An essential feature of locked-in patients' condition is that they can no longer express their emotions through body language and gestures. We now know that the muscle activity and changes in tension necessary for such emotional expression are carefully registered and processed. A smile or a frown are not only the products of feelings, they also produce emotions themselves. This usually means they reinforce feelings that are already there. When people are angry, they make an angry face, and that in turn reinforces their ire. That mechanism no longer works in locked-in patients, and so their emotions become less extreme. At both the positive and the negative end of the scale. Neither anger nor joy have a tendency to become extreme; everything tends towards equanimity. A phenomenon that seems to echo a famous saying from Gautama Buddha:

'The wise show no elation or depression when touched by happiness or sorrow.'

However, as we know, the principle feature of the locked-in state is that the brain no longer has any means of achieving effects. We are naturally shocked when we think about motorcyclists or horse riders waking up completely paralysed after an accident. Such a loss of the ability to move is so sudden that the person in question has no time to process and come to terms with it. But a gradually encroaching paralysis, as with ALS or Parkinson's disease, appears particularly horrific to us. The British historian and essayist Tony Judt, who contracted ALS, describes his slide into isolation as an unbearable feeling that everything is gone:

'Gone is the yellow pad, with its now useless pencil. Gone the refreshing walk in the park or workout in the gym, where ideas and sequences fall into place as if by natural selection. Gone too are productive exchanges with close friends — even at the midpoint of decline from ALS, the victim is usually thinking far faster than he can form words, so that the conversation itself becomes partial, frustrating and ultimately self-defeating.'[2]

Another ALS patient speaks of feeling like the 'undead', avoided by an indifferent or even disgusted environment: 'I have even become useless and boring to my cat since she realised I can't stroke her anymore.' Others find it particularly difficult to cope with the constant emotional rollercoaster of joy and sadness, hope and resignation, optimism and gloom that accompanies the disease.

These are all descriptions of complete paralysis creeping into people's lives, destroying everything that was once important to them. It is clear that such a process cannot generate positive feelings. But what happens when the process is complete? When all hope of improvement is finally extinguished, when no movement is possible and virtually all vital processes are dependent on machines?

All the signs are that desperation also disappears and is replaced by a soothing emptiness.

When the brain can no longer achieve effects, it can also no longer experience failure. And it can no longer feel fear, since it now has no means of triggering a fight-or-flight reaction. So, locked-in patients are spared those negative experiences. It is no coincidence that we found next to no activity in their brains' defence systems.

Equally important for the positive feeling of emptiness, however, is that the will in general ebbs away, which was postulated by Buddhism and by Schopenhauer as the essential prerequisite for liberation from the cycle of suffering. When the brain can no longer achieve anything, it will eventually stop doing anything to try. This means that, as scientists, we in Tübingen often have to use all our powers of persuasion and technical tricks — such as waking patients after we see EEG patterns indicative of falling asleep — to drag our locked-in patients out of their inactivity so that we can communicate with them. After all, we cannot expect them to be ready to work with us permanently and at all times. Indeed, there are more fascinating pastimes than repeatedly answering questions like 'Do you like it when your family comes to visit?' or 'Do elephants have beaks?' For the patients themselves, the ebbing away of their will is not a drama. On the contrary, the data we have gathered so far lead us to believe that they experience it as a liberation from the recurring dramas that our will tends to force upon us daily.

It is not only that locked-in patients no longer feel disappointment or loss. They also no longer feel any urges or desires, no yearnings or cravings. They are able to observe things objectively and without bias, because they invest no hope in them. That said, brain scans of our patients have indicated activity when a positive event was in the offing, such as a visit from the patient's children. This means they must still harbour hopes in connection with the anticipation of events — but not in connection with actions, since they are no longer able to perform any (at least, we must assume this, on the basis of the physiological changes we have measured in their brains). In that sense, they are empty, and that means that they view the world without any subjective desires.

It is possible that locked-in patients view the world much more objectively, penetrating and understanding it far more than we could ever do. Does that still sound like a dungeon or eternal damnation? It is more reminiscent of that disinterested and unadulterated dimension of enlightenment that some philosophers strive for as the source of a happiness truly free of desire. It is no coincidence that one Zen saying goes: 'A life without hope is a life full of peace, joy, and compassion.' This is the state locked-in patients eventually enter.

Silence, not screen-time

Of course, many family members and carers see this differently. The vast majority of the time, locked-in patients are not hooked up to machines that enable them to communicate. They lie in their beds, and the only movement in the room is the tireless, hissing motion of the ventilation machine. It is more reminiscent of rigor mortis than of life, and it is understandably a sight that family members find difficult to bear. Patients are often placed in front of a television set for long periods of time, so that they're not quite so cut off from the outside world, and to provide them with some stimulation. Just as people do with their pet parrots when they have to leave them at home alone for a longer period of time.

Only, locked-in patients are not parrots. As soon as they're asked to say, via neurofeedback, whether they would like the television to be turned off, they almost always answer with a clear 'yes'. When asked if they like having the TV running the whole time, they usually say 'no'. This leads us to suspect that, while locked-in patients will occasionally communicate with their environment, they mostly want to be left alone in their emptiness. Most importantly, they do not want to be exposed to moving images. Such images are simply too sharp a contrast to that which they are now experiencing: stillness and emptiness. Why should a person who has lost all motor abilities and can't even bat an eyelid be interested in moving pictures? It is comparable to gushing to a deaf person about Beethoven's Ninth, or showing Gaugin's brightly coloured Tahiti paintings to a blind person. In fact, we must assume that locked-in patients can no longer relate to movement in any way and are basically unable to perceive it or process it mentally.

This explains why, when we confronted our patients with the usual questions and statements about happiness and satisfaction with life — such as 'Do you enjoy meeting with friends?' or 'I like getting up in the morning'

— we very rarely registered positive reactions. Both of those activities are impossible for a completely paralysed person. When they were presented with true-or-false statements of fact, however, such as 'I have good friends', we got a much greater response. They reacted far more positively to the statement 'Life is beautiful' than they did to 'I enjoy life'.

Waltraut Fähnrich's husband, too, often receives no response or a disappointing 'no' when his questions or statements are too dynamic in nature or refer to a too-distant future prospect. He once asked his wife if she was looking forward to celebrating their golden wedding anniversary with him in three years' time. He was met with rebuff, a clear 'no', despite her frequently professed affection for him. This was because, for someone whose life has no movement and therefore no development, questions referring to three years in the future simply have no meaning.

Does that mean, then, that locked-in patients conceive the world as static? When we think, when we dream, our thoughts and dreams are usually dynamic because the principle activity our brain can instigate is motion, and that means movement forms the basis of our daily experience: A does something, B reacts to A, or at least could potentially react to it, and usually we take the place of either A or B. Locked-in patients have lost that pattern. What happens, then, to the majority of their thoughts? The fact that they stop responding to questions referring to activity is already a sign that motion no longer plays a particular part in their thought-world. What remains?

Just being rather than functioning

In order to answer that question, we conducted a study in which we confronted our patients either with verbs, such as 'eat', 'walk', 'love', or 'think', or with nouns, such as 'meal', 'chair', 'affection', or 'thought'. From tests of healthy subjects, we know that the associative network for verbs and movement-related linguistic constructions is located in the central regions of the left hemisphere, because verbs produce high-frequency — between 30 and 40 Hz — brainwaves there. Nouns, by contrast, are thought about in the posterior sensory areas. Our plan was now to present our locked-in subjects with 60 verbs and 60 nouns, and their tasks included indicating, by thinking 'yes' or 'no' and triggering the corresponding changes in brain activity, whether they would classify each word as a noun or a verb, and whether they would categorise it as emotionally positive, negative, or neutral.

Our hypothesis — developed on the basis of previous observations — was that the verbs would *not* trigger great activity in the corresponding areas of the brain in the way they would in the brains of healthy subjects. The static nouns, on the other hand, should produce similar brain activity in both locked-in patients and healthy subjects.

In addition, we planned to read stories aloud to the patients and then require them to indicate their recall of the tales by playing them words that either corresponded with the content of the story or didn't. They had to react with a 'yes' or a 'no', and the answer was registered using near-infrared technology. We suspected that the locked-in patients would mainly list nouns, and that verbs would be under-represented.

The results of this study are not yet ready for publishing, but it is worth considering already what it would mean if our hypothesis is confirmed and locked-in patients really do think more in terms of nouns, and thoughts about motion and action no longer play a significant part in their mental world. What must that be like? Something akin to looking at pictures in a photo album? Or perhaps even something like the flash-freezing technology used in the food industry, which instantly transforms elastic tissue into a solid, deep-frozen block?

The problem is that we are unable to imagine thinking only in terms of nouns as anything other than fixed and static, because we are dynamic, motor-oriented beings. But if we go back to the concept of emptiness in Zen Buddhism, we see that the precise opposite may be true. In Zen, when things lose their dynamic quality and their function, they cease to act and therefore also to react *with each other*. The bird no longer perches on the branch, the branch no longer grows from the tree, the clouds no longer scud across the sky, and rain no longer falls from the clouds. Things present themselves as simply being, and not in relation to one another. And then the borders between them become blurred, so to speak, as does the distinction between the observer and the observed. 'The bird regards the flower and the flower regards the bird,' says the Zen master Dōgen, 'The bird is the flower and the flower is the bird.' The battle between opposites, the old temple founder went on, 'gives way to all-encompassing kindness.'

It may seem strange to compare the state of locked-in patients to the emptiness of Zen, and we do not have any real proof that the two are genuinely similar. But there is good reason to continue to think along those lines and to do further research. We are certainly not saying that everyone should be given the opportunity to enter a state of emptiness — with drugs, for example — because they will find absolute bliss there. There is still too little scientific knowledge to prove that hypothesis. But we are calling for a change of perspective. As creatures endowed with a will, we should at least make an attempt to understand the world of almost will-less locked-in patients. Without any preconceived ideas! And part of that means not automatically calling for them to be 'switched off' and given a 'dignified end' when we are confronted with the fate of such people.

Emptiness as the Beginning of the End of Life

how emptiness will return to us at last

'When I die, I want to see the North Pole before me. That emptying of everything, all colour gone, all nature, only white-grey, mouse-grey, a light shimmer of gold. That would be an end. A good end.'

These are the words of the German author and TV presenter Roger Willemsen, who died in February 2016. They appeared in an interview published in the magazine *Emotion* in 2011. Clearly, this eloquent, intellectual man had some very concrete wishes when it came to his own death. He wanted it to be an 'emptying', free of everything that normally makes up life. Grey, with a little shimmer of gold, and then, the end. It's clear Willemsen did not think of emptiness as something negative, but rather as full of promise. No one knows whether his experience of death lived up to that promise. But it is possible. There are studies that indicate that dying could really be a kind of emptying. Life's final emptying.

Staring at the top of the doctor's bald head

Since time immemorial, people have speculated about what happens when life expires. This includes both the question of the afterlife, and what happens during death. There can be no certain knowledge about this. After all, death is final and no one has ever come back from the dead to report on it. But we do at least have some clues as to what it might be like.

There are relatively abundant reports from people who came pretty close to dying. Around 10 per cent of heart-attack survivors report having had a so-called near-death experience. We hear of a bright light guiding the dying person down a tunnel, or of out-of-body experiences. Near-death experiences are generally described as particularly vivid and clear, but at the same time extremely calm and free of fear. Many scientists believe these are nothing more than hallucinations, arguing that a brain starved of blood and nutrients is no longer able to execute any coordinated operations. But recent studies have cast doubt on this view.

A team of researchers at the University of Michigan in Ann Arbor used EEG to analyse the brainwaves of anaesthetised rats while they were undergoing a heart attack induced with an injection of a potassium-chloride solution.[1] They found that in the first four seconds the overall amplitude of the rats' brainwaves fell sharply and then stabilised at a low level for approximately the next six seconds. This seems to support the theory that dying is like simply pulling the plug on the brain, leaving it unable to execute coordinated processes.

But then something astonishing happened: for 30 seconds, the EEG readings showed the development of a gently undulating sea of alpha and theta waves out of which rose pronounced, synchronised gamma oscillations. This wave pattern is similar to that found in the brains of advanced meditation practitioners and professional musicians. What we don't know is whether the gamma activity in rats is centred on the previously described salience network of the anterior insula and anterior cingulate cortex. Nevertheless, it seems the beginning of death is connected with intense, subjective experience, which means that great care must be taken when declaring someone dead! What is certain is that the dying rodents were profoundly relaxed on the one hand, but displayed islands of intense, coordinated brain activity on the other. There was no sign of fear, or of simply drifting away.

When applied to humans, these results confirm those of a study carried out by the Stony Brook Medical Centre in New York.[2] The American researchers chose 140 cardiac-arrest survivors who had suffered temporary cardiac death, and interviewed them soon after their resuscitation. The interviews were designed to elicit information about near-death experiences. Prior to the beginning of the study, the researchers cleverly placed various colour images in the rooms where resuscitations usually took place. The images were placed on shelves in such a way that they could only be seen from above. This was to test reports of out-of-body experiences during cardiac death.

Fifty-five patients in total reported having memories of the time between their cardiac arrest and their successful reanimation. Around half reported having the impression that everything happened faster or slower than usual during that time. Twenty-two reported having a feeling of peace or pleasantness, and 13 said they felt their senses were more vivid than usual, although they could not remember any details of what they had perceived. This is interpreted by Sam Parnia, the lead scientist in the study, as a sign that 'patients may have greater conscious activity during cardiac arrest than is evident through explicit recall, perhaps due to the impact of post-resuscitation global cerebral inflammation and/or sedatives on memory consolidation and recall'.

Another possible interpretation is that awareness during a near-death experience is indifferent, as is the case with meditation, and so does not leave anything behind that might be memorable. It is no coincidence that considerable memory gaps have also been recorded in meditation practitioners, not only holes in their memory of a meditation session, but also with respect to other events. Someone who has been trained to see things in terms of their meaninglessness will store fewer memories — because only things that have meaning are given a place in our mental memory banks.

Thirteen patients in the study reported feeling separated from their body during their cardiac arrest.

Unfortunately, they were the very ones who happened not to be in the rooms where the pictures had been positioned on the high shelves. However, the recollections of one 57-year-old patient were remarkable even without the shelf-pictures. In his interview, he reported watching part of his resuscitation from above. He told of seeing a man and a nurse working on his body: 'I couldn't see his face but I could see the back of his body. He was quite a chunky fella … He had blue scrubs on, and he had a blue hat, but I could tell he didn't have any hair, because of where the hat was.' He also recalled hearing an automated voice saying, 'shock the patient, shock the patient'. The next thing he remembered was waking up and hearing the nurse saying, 'Oh you nodded off … you are back with us now.'

The scientists were able to confirm that an automatic defibrillator with a computer-simulated voice function had been used during the man's resuscitation. It was also true that the reanimation had been carried out by a stout doctor with a bald head, which would not have been visible from the front due to the position of his headgear. 'He accurately described people, sounds, and activities from his resuscitation,' stresses Parnia. 'His medical records corroborated his accounts and specifically supported his descriptions.'

Emptiness knows no ifs and buts

We need not speculate further on whether near-dead people can really float in the air. But it is useful to examine the sequences of events described by people during their near-death experiences. One thing to note is that none ever mention a feeling of fear, and many describe a sense of 'profound peace'.

This tallies with near-death experiences recounted by mountaineers, for example, who have fallen from great heights and were convinced that they were experiencing the final few moments of their lives. They also report feeling no fear. They describe, more frequently than resuscitated cardiac patients, seeing their whole lives pass before them like a movie. This may have to do with the fact that they were not lying prone on a bed, but hurtling earthward at great speed. On the other hand, it may well be that the dying lab rats from the study described above also saw their whole lives flash before them, as their brain activity in the high-frequency gamma range could be a sign of the hippocampus vigorously fetching memories from its store one last time.

Whatever the case, it seems that all near-death experiences occur without fear and create highly active but indifferent islands of awareness in a sea of relaxation, which we know from Chapter 7 is also a typical feature of deep meditation. The out-of-body experiences described by many who have been close to death also chime with the meditative experience of emptiness: the boundaries of the self are transcended, as in descriptions of mystical religious experiences. A full 8 per cent of interviewees in the near-death study even described drifting into some other, unearthly world; 3 per cent spoke of a 'feeling of joy'.

It seems there is no need for us to worry inordinately about dying. As long as no fear-inducing factors such as respiratory distress or pain are involved — which should not be the case in this day and age — the great emptying at the end of our lives seems to generate a relaxed, and in some cases even euphoric mood. Of course, there can be no certainty about this.

It also raises the question of why the final experience of emptiness in particular is often felt to be pleasant. It certainly is an odd idea that hospital doctors should be swarming around a patient, fighting for his or her life, while he or she experiences something akin to an orgasm, or at least to meditation. Why does desperation not set in? After all, it's the patient's life, and not the doctor's, which is in the process of ending.

The answer is that a dying person has accepted the inescapability of imminent death and has no more hope of being able to do anything to change the situation. This robs fear of its meaning, the defence system comes to rest, and, with that, all thoughts connected with fear cease to circulate. The field of psychology teaches us that phobics can be cured of their fears if they are confronted with the situation they find frightening in a context where there is no opportunity to flee.

When I jumped out of that plane, my panic switched off at the precise moment when I suddenly realised that there was nothing I could do now to change anything — allowing me to fall not only towards the Earth, but also into a soothing emptiness. Just like those mountaineers, when they fall and are convinced they are plummeting to their certain death. The philosopher Emil Cioran told how he was filled with 'the ecstasy of capitulation' after he had finally accepted his insomnia as an irreversible fate. Although he was still unable to sleep, his desperation was gone and he stared into an 'abyss without vertigo'. Epileptics also report that struggling to ward off an imminent seizure does nothing to improve the situation. This is why they willingly let themselves be swept along by their fits; some even trigger them on purpose. ALS patients, too, are desperate at first, when they are still struggling against the merciless onward march of their disease without any way to stop the process that they know will leave them increasingly trapped in their own bodies. But when they are eventually completely locked-in, they seem not only to come to terms with emptiness, but even to enjoy an improved quality of life.

In order to achieve a positive experience of emptiness, it is therefore necessary not to focus on what is lost through that achievement. And it is necessary for there to be no back door through which it might be possible to escape the journey towards emptiness if it should become unpleasant. It is necessary to abandon control of oneself and drift towards nothingness with no ifs or buts.

Back to the womb

Perhaps emptiness can offer us more, however, than just the ebbing away of our defence system and the cessation of the merry-go-round of thoughts that accompany it. In fact, emptiness propels us back to the time before we were born, and still experiencing something similar to Peter Suedfeld's floatation tanks, inside our mother's womb: floating in an environment of sensory deprivation.

Everything is dark and more or less quiet, and our sense of proprioception also has little to report when we are surrounded by amniotic fluid.

We have been able to observe that the brains of unborn babies in the final three months of gestation predominantly produce low-frequency waves typical of the 'twilight state'. Physiologists used to believe that foetuses dream a lot during this phase, but they were at a loss to say what an unborn baby might dream about before having gathered any experiences. Twilight brainwaves make much more sense here: they enable the baby to survive the locked-in situation in the womb. This means the capacity to achieve a state of emptiness is pre-programmed in us before we are born. And so it should surprise no one that we retain an affinity for it throughout our lives. Indeed, the more surprising fact is that we repeatedly fear emptiness. Yet, as we have seen, that is due to our modern, content-driven lifestyle, and to diseases like depression or borderline personality disorder, in which emptiness is seen as a threat.

In the womb, by contrast, it is the end of emptiness that is the threat. After nine months, the times comes for the foetus to leave its warm and cosy cave and be squeezed towards a blinding light. It is not for nothing that Sigmund Freud saw birth as a traumatic experience. We now know, however, that so many endorphins are flowing through the baby's system at the time that the trauma gives way to pleasure. It is striking how similar the process of birth is to recollections of near-death experiences in which people describe being drawn through a tunnel towards a light.

Is that a coincidence? Or do the experience of death and birth really have something in common?

For Arthur Schopenhauer, there was no doubt about it. He believed birth is the way out of nothingness, and death is the way back to nothingness. Death returns us to where we came from: 'The condition to which death returns us is our original one, that is, the very condition of being.' And, according to Schopenhauer, this condition is one in which the will is no longer something that enters into 'wanting subjects', but remains completely in and of itself, thus no longer causing any suffering that must be endured by those people. Just as air is still air when there is no one who needs to breathe it in order to stay alive. Schopenhauer concludes from this that it is foolish to fear death. Since with death, the suffering 'subject of the will' also disappears, although the will itself remains and can now exist for itself without a subject.

Here, Schopenhauer enters the realm of metaphysics, which for many people is as difficult to fathom as the fate of locked-in patients, because in both cases the 'wanting subject' is extinguished. But the great philosopher of release is probably right again. Whatever the case, when we are born, we leave emptiness behind us and we must work hard during our lives to regain it, at least temporarily.

As we have seen, this work includes inducing certain brain processes. Most importantly, the brain's defence system, centred on the amygdala, must come to rest; connections to the systems of the will in the basal ganglia must also be interrupted. There are many ways to achieve this: brain stimulation or meditation, marching in step or staring out to sea, chanting in the football stadium or floating in a tank, having orgasms or jumping out of planes. We have described some of these ways and tried to trace the processes in the human brain which are involved. We would like readers to take this as an encouragement to make their own attempts.

But readers should not expect too much. Neither a eureka moment nor enlightenment, neither ecstasy nor total consciousness, neither freedom nor liberation. Such expectations would simply activate the will and corresponding systems in the brain and thereby banish emptiness. Emptiness cannot be willed. Quite the contrary! The more energetically we strive for it, the more it slips out of our grasp.

So, what *can* we do? We can behave as if emptiness will come to us of its own accord, since it opens itself up only to those who place no hope in it.

Notes

Introduction

1   Sagioglou, C. & Greitemeyer, T., 'Individual Differences
    in Bitter Taste Preferences Are Associated with Antisocial
    Personality Traits', *Appetite* 96(5–6), 2016

Chapter 1

1   Schulze, G., *Die Erlebnisgesellschaft. Kultursoziologie der
    Gegenwart*, Frankfurt 1992, published in 1995 in English
    as *The Experience Society*

2   Bausinger, H., *Ergebnisgesellschaft. Facetten der Alltagskultur*,
    Tübingen 2015

3   Wilson, T.D. et al., 'Just Think: the challenges of a disengaged
    mind', *Science* 345(6192), 2014

Chapter 2

1   App, U. (ed.), *From the Record of the Chan Master 'Gate of
    the Clouds'*, Bern 1994

2   Suzuki, D.T., 'An Interpretation of Zen Experience' in
     Studies in Zen, London 1955

3   cf. Han, B.-C., *Philosophie des Zen-Buddhismus*, Stuttgart 2002

4   Würzbach, F., *Nietzsche. Sein Leben in Selbstzeugnissen,
     Briefen und Berichten*, Munich 1968

5   Lange-Eichbaum, W., *Nietzsche. Krankheit und Wirkung*,
     Hamburg 1946

6   Sax, L., 'What Was the Cause of Nietzsche's Dementia?',
     Journal of Medical Biography 11(1), 2003

Chapter 3

1   Carlson, N.R. et al., *Psychology: the science of behaviour*,
     New York 2013

Chapter 4

1   Feinstein, J. et al., 'The Human Amygdala and the Induction
     and Experience of Fear', *Current Biology* 21(1), 2011

2   Moutsiana, C. et al., 'Insecure Attachment during Infancy
     Predicts Greater Amygdala Volumes in Early Adulthood',
     Journal of Child Psychology and Psychiatry 56(5), 2015

3   LeDoux, J., *The Emotional Brain: the mysterious underpinning
     of emotional life*, New York 1996

4   Lyons, I.M. et al., 'When Math Hurts: math anxiety predicts
     pain network activation in anticipation of doing math', *PLoS
     ONE* 7(10), 2012

Chapter 5

1   Raichle, M. et al., 'A Default Mode of Brain Function', *Proceedings of the National Academy of Sciences of the USA* 98(2), 2001

2   Christoff, K. et al., 'Experience Sampling during fMRI Reveals Default-network and Executive-system Contributions to Mind-wandering', *Proceedings of the National Academy of Sciences of the USA* 106(21), 2009

3   Buckner, R.L. et al., 'The Brain's Default Network: anatomy, function, and relevance to disease', *Annals of the New York Academy of Sciences* 1124, 2008

4   Schilbach, L. et al., 'Minds at Rest?: social cognition as the default mode of cognising and its putative relationship to the "default system" of the brain', *Consciousness and Cognition* 17(2), 2008

5   Harrison, B.J. et al., 'Consistency and Functional Specialisation in the Default-mode Brain Network', *Proceedings of the National Academy of Sciences of the USA* 105(28), 2008

6   Killingsworth, M. et al., 'A Wandering Mind is an Unhappy Mind', *Science* 330(11), 2010

7   Birbaumer, N., 'Breaking the Silence: brain–computer interfaces (BCI) for communication and motor control', *Psychophysiology* 43(6), 2006

Chapter 6

1   cf. Brown, R. & Milner P.M., 'The Legacy of Donald O. Hebb: more than the Hebb Synapse', *Nature Reviews Neuroscience* 4, 2003

2   Lilly, J., *The Deep Self: consciousness exploration in the isolation tank*, London 2006

3   Suedfeld, P. (ed.), *Light from the Ashes: social science careers of young Holocaust refugees and survivors*, Ann Arbor 2001

4   Suedfeld, P., *Restricted Environmental Stimulation: research and clinical applications*, New York 1980

5   Suedfeld, P. et al., 'Autobiographical Memory and Affect under Conditions of Reduced Environmental Stimulation', *Journal of Environmental Psychology* 15(4), 1995

6   Bell, C., *The Hand: its mechanism and vital endowments as evincing design*, Bridgewater 1833

7   Head, H. et al., *Studies in Neurology* vol. 2, London 1920

8   Cole, J. & Waterman, I., *Pride and a Daily Marathon*, Cambridge 1995

9   Tsakiris, M. et al., 'My Body in the Brain: a neurocognitive model of body-ownership', *Neuropsychologia* 48(3), 2010 Chapter 7

Chapter 7

1   Naqvi, N. et al., 'Damage to the Insula Disrupts Addiction to Cigarette Smoking', *Science* 315(5811), 2007

2   Van de Wetering, J., *The Empty Mirror: experiences in a Japanese Zen monastery*, New York 1973

3   Newberg, A. et al., *Why God Won't Go Away: brain science and the biology of belief*, New York 2002

Chapter 8

1   Bianchi-Demicheli, F. et al., 'Neural Bases of Hypoactive Sexual Desire Disorder in Women: an event-related FMRI study', *The Journal of Sex Medicine* 8(9), 2011

2 Chuang Y.C. et al., 'Tooth-brushing Epilepsy with Ictal Orgasms', *Seizure* 13(3), 2004

3 Dluzen, D. & Ramirez, V., 'Measurement of Hormonal and Neural Correlates of Reproductive Behaviour', *Methods of Neurosciences* 14, 1993

4 Holstege, G. et al., 'Brain Activation during Human Male Ejaculation', *The Journal of Neuroscience*, 23(27), 2003

5 Abramson, P. & Pinkerton, S., *With Pleasure: thoughts on the nature of human sexuality*, New York/Oxford, 1995

6 Maloy, K. & Davis, J., '"Forgettable" Sex: a case of transient global amnesia presenting to the Emergency Department', *The Journal of Emergency Medicine* 41(3), 2011

7 Ramachandran, V. et al., *Phantoms in the Brain: probing the mysteries of the human mind*, New York, 1998

8 Cook, C.M. & Persinger, M., 'Experimental Induction of the "Sensed Presence" in Normal Subjects and an Exceptional Subject', *Perceptual and Motor Skills* 85(2), 1997

9 Bragagna, E., 'Der weibliche Orgasmus: Jenseits von Mythen', *gynäkologie + geburtshilfe*, 2013/S1

Chapter 9

1 Winkler, I. et al., 'Newborn Infants Detect the Beat in Music', *Proceedings of the National Academy of Sciences of the USA* 106(7), 2009

2 Zentner, M. & Eerola, T., 'Rhythmic Engagement with Music in Infancy', *Proceedings of the National Academy of Sciences of the USA* 107(13), 2010

3   Hove, M. et al., 'Superior Time Perception for Lower Musical
     Pitch Explains Why Bass-ranged Instruments Lay Down
     Musical Rhythms', *Proceedings of the National Academy of
     Sciences of the USA* 111(28), 2014

4   Cameron, D.J. et al., 'Cross-cultural Influences on Rhythm
     Processing: reproduction, discrimination, and beat tapping',
     Frontiers in Psychology 6(366), 2015

5   Harrison, L. & Loui, P., 'Thrills, Chills, Frissons, and Skin
     Orgasms: toward an integrative model of transcendent
     psychophysiological experiences in music', *Frontiers in
     Psychology* 5, 2014

6   www.youtube.com/playlist?list=PLdCNGoNr0dbnS5kKe-
     hbJms1jWN-QhXdf

Chapter 10

1   Wallace, D.F., *Infinite Jest*, Little Brown, 1996

2   Schmaal, L. et al., 'Subcortical Brain Alterations in Major
     Depressive Disorder: findings from the ENIGMA Major
     Depressive Disorder working group', *Molecular Psychiatry* 21, 2016

3   Andrews, P. et al., '*Primum non nocere*: an evolutionary analysis
     of whether antidepressants do more harm than good', *Frontiers
     in Psychology* 3(117), 2012

4   von Helversen, B. et al., 'Performance Benefits of Depression:
     sequential decision-making in a healthy sample and a clinically
     depressed sample', *Journal of Abnormal Psychology* 120(4), 2011

5   Goenner, H., *Albert Einstein*, Munich, 2015

6   Raine, A., 'Autonomic Nervous System Factors Underlying
     Disinhibited, Antisocial, and Violent Behaviour: biosocial

perspectives and treatment implications', *Annals of the New York Academy of Sciences* 794, 1996

7 Dziobek, I. et al., 'Neuronal Correlates of Altered Empathy and Social Cognition in Borderline Personality Disorder', *NeuroImage*, 57(2), 2011

8 Bleuler, E., *Allgemeine Zeitschrift für Psychiatrie und psychischgerichtliche Medizin* 65, 1908

9 van Erp, T.G.M. et al., 'Subcortical Brain Volume Abnormalities in 2028 Individuals with Schizophrenia and 2540 Healthy Controls via the ENIGMA Consortium', *Molecular Psychiatry* 21, 2016

10 Henneman, W.J.P. et al., 'Hippocampal Atrophy Rates in Alzheimer Disease', *Neurology*, 72(11), 2009

11 Max Planck Institute for Human Development, 21 October 2014

Chapter 11

1 This incisive formulation is taken from the science writer Ernst Peter Fischer; he used it in his science blog, in a post about our studies of locked-in patients, and it is an oblique reference to Theodor Adorno's well-known phrase (from his book *Minima Moralia*), 'Wrong life cannot be lived rightly.'

2 Judt, T., *The Memory Chalet*, London, 2010

Chapter 12

1 Borjigin, J. et al., 'Surge of Neurophysiological Coherence and Connectivity in the Dying Brain', *Proceedings of the National Academy of Sciences of the USA* 110(35), 2013

2 Parnia, S. et al., 'AWARE — AWAreness during REsuscitation: a prospective study', *Resuscitation* 85(12), 2014